NATIONAL ACADEMIES

Sciences
Engineering
Medicine

NATIONAL
ACADEMIES
PRESS
Washington, DC

Advancing Risk Communication with Decision-Makers for Extreme Tropical Cyclones and Other Atypical Climate Events

Laura Yoder and John Ben Soileau,
Rapporteurs

Board on Atmospheric Sciences and
Climate

Board on Environmental Change and
Society

Division on Earth and Life Studies

Division of Behavioral and Social
Sciences and Education

Proceedings of a Workshop

NATIONAL ACADEMIES PRESS 500 Fifth Street, NW Washington, DC 20001

This activity was supported by a contract between the National Academy of Sciences and the National Aeronautics and Space Administration, National Oceanic and Atmospheric Administration, and National Science Foundation. Any opinions, findings, conclusions, or recommendations expressed in this publication do not necessarily reflect the views of any organization or agency that provided support for the project.

International Standard Book Number-13: 978-0-309-72537-8
International Standard Book Number-10: 0-309-72537-2
Digital Object Identifier: https://doi.org/10.17226/27933

This publication is available from the National Academies Press, 500 Fifth Street, NW, Keck 360, Washington, DC 20001; (800) 624-6242 or (202) 334-3313; http://www.nap.edu.

Suggested citation: National Academies of Sciences, Engineering, and Medicine. 2025. *Advancing Risk Communication with Decision-Makers for Extreme Tropical Cyclones and Other Atypical Climate Events.* Washington, DC: The National Academies Press. https://doi.org/10.17226/27933.

The **National Academy of Sciences** was established in 1863 by an Act of Congress, signed by President Lincoln, as a private, nongovernmental institution to advise the nation on issues related to science and technology. Members are elected by their peers for outstanding contributions to research. Dr. Marcia *McNutt* is president.

The **National Academy of Engineering** was established in 1964 under the charter of the National Academy of Sciences to bring the practices of engineering to advising the nation. Members are elected by their peers for extraordinary contributions to engineering. Dr. John L. Anderson is president.

The **National Academy of Medicine** (formerly the Institute of Medicine) was established in 1970 under the charter of the National Academy of Sciences to advise the nation on medical and health issues. Members are elected by their peers for distinguished contributions to medicine and health. Dr. Victor J. Dzau is president.

The three Academies work together as the **National Academies of Sciences, Engineering, and Medicine** to provide independent, objective analysis and advice to the nation and conduct other activities to solve complex problems and inform public policy decisions. The National Academies also encourage education and research, recognize outstanding contributions to knowledge, and increase public understanding in matters of science, engineering, and medicine.

Learn more about the National Academies of Sciences, Engineering, and Medicine at **www.nationalacademies.org**.

Consensus Study Reports published by the National Academies of Sciences, Engineering, and Medicine document the evidence-based consensus on the study's statement of task by an authoring committee of experts. Reports typically include findings, conclusions, and recommendations based on information gathered by the committee and the committee's deliberations. Each report has been subjected to a rigorous and independent peer-review process and it represents the position of the National Academies on the statement of task.

Proceedings published by the National Academies of Sciences, Engineering, and Medicine chronicle the presentations and discussions at a workshop, symposium, or other event convened by the National Academies. The statements and opinions contained in proceedings are those of the participants and are not endorsed by other participants, the planning committee, or the National Academies.

Rapid Expert Consultations published by the National Academies of Sciences, Engineering, and Medicine are authored by subject-matter experts on narrowly focused topics that can be supported by a body of evidence. The discussions contained in rapid expert consultations are considered those of the authors and do not contain policy recommendations. Rapid expert consultations are reviewed by the institution before release.

For information about other products and activities of the National Academies, please visit www.nationalacademies.org/about/whatwedo.

WORKSHOP PLANNING COMMITTEE

ANN BOSTROM (*Chair*), Weyerhaeuser endowed Professor in Environmental Policy at the Evans School of Public Policy and Governance, University of Washington

DEREKA CARROLL-SMITH, Postdoctoral Research Associate, National Institute of Standards and Technology Professional Research Experience Program, University of Maryland College Park

BRAD R. COLMAN, President, American Meteorological Society

WILLIAM C. FUGATE, Principal, Craig Fugate Consulting LLC.

MICHAEL K. LINDELL, Emeritus Professor, Texas A&M University; Affiliate Professor, University of Washington Department of Urban Design and Planning; Affiliate Professor, Boise State University Department of Geosciences; and Affiliate Professor, Oregon State University School of Civil and Construction Engineering

ANDREA B. SCHUMACHER, Project Scientist, Weather Risks and Decisions in Society research group, National Center for Atmospheric Research

J. MARSHALL SHEPHERD, Georgia Athletic Association Distinguished Professor of Geography and Atmospheric Sciences and Director of Atmospheric Sciences Program, University of Georgia

JEANNETTE SUTTON, Associate Professor, Department of Emergency Preparedness and Homeland Security and Director of Emergency and Risk Communication Message Testing Lab, University at Albany

National Academies of Sciences, Engineering, and Medicine Staff

JOHN BEN SOILEAU, Program Officer
HUGH WALPOLE, Associate Program Officer (until March 2024)
KATELYN CREWS, Program Assistant (from March 2024)
RITA GASKINS, Administrative Coordinator (until March 2024)
ROB GREENWAY, Program Associate (until March 2024)
MAGGIE WALSER, Acting Co-Director, Board on Atmospheric Sciences and Climate

Reviewers

This Proceedings of a Workshop was reviewed in draft form by individuals chosen for their diverse perspectives and technical expertise. The purpose of this independent review is to provide candid and critical comments that will assist the National Academies of Sciences, Engineering, and Medicine in making each published proceedings as sound as possible and to ensure that it meets the institutional standards for quality, objectivity, evidence, and responsiveness to the charge. The review comments and draft manuscript remain confidential to protect the integrity of the process.

We thank the following individuals for their review of this proceedings:

ANN BOSTROM, University of Washington
DAPHNE LADUE, University of Oklahoma
KELLY MCCUSKER, Rhodium Group

Although the reviewers listed above provided many constructive comments and suggestions, they were not asked to endorse the content of the proceedings nor did they see the final draft before its release. The review of this proceedings was overseen by **Edwin Schmitt**, California Public Utilities Commission. He was responsible for making certain that an independent examination of this proceedings was carried out in accordance with standards of the National Academies and that all review comments were carefully considered. Responsibility for the final content rests entirely with the rapporteur(s) and the National Academies.

Contents

Boxes and Figures

BOXES

FIGURES

Acronyms and Abbreviations

AI	artificial intelligence
AMS	American Meteorological Society
CAP	common alerting protocol
CAT	Communication Assist Tech
CMAC	Commercial Mobile Alert for C-Interface
DBGF	Device-Based Geofencing
EAS	Emergency Alert System
EM	emergency manager
EOC	Emergency Operations Center
FEMA	Federal Emergency Management Agency
GIS	Geographic Information Systems
ICE	U.S. Immigration and Customs Enforcement
IDSS	Impact-Based Decision Support Services
IPAWS	Integrated Public Alert System
IPCC	Intergovernmental Panel on Climate Change
NASA	National Aeronautics and Space Administration
NCAR	National Center for Atmospheric Research
NHC	National Hurricane Center
NOAA	National Oceanic and Atmospheric Administration
NSF	National Science Foundation
NWS	National Weather Service
SDII	Societal Data Insights Initiative
SPARK	System for Public Access to Research Knowledge
SUNY	State University of New York
TORFF	Tornadoes and Flash Floods
U.S.	United States
USGS	United States Geological Survey
WEA	Wireless Emergency Alerts
WPO	Weather Program Office

Chapter 1

Introduction

Extreme weather events (e.g., tropical cyclones, major winter storms, wildfires) pose substantial threats to life, property, and livelihoods in the United States and worldwide. Over the past several decades, major advances in forecasting capabilities have enabled the creation of more accurate, detailed, and nuanced risk information products. Despite these advances, communicating about extreme weather events with decision-makers and the public remains challenging but opportunities for continued innovation exist.

Decision-making related to extreme weather events often involves tradeoffs between different degrees and types of risks. In the context of tropical cyclones, for example, deciding whether or when to issue an evacuation order entails weighing the risks to lives posed by the event against the risks to livelihoods posed by the financial costs of evacuations or relocations. The inherent uncertainty in extreme events complicates the calculation of such tradeoffs. The tradeoffs may be obvious when the timing, location, and magnitude of impacts are known with high certainty, but this level of certainty is rarely, if ever, the case for extreme weather events. Rather, despite dramatic improvements in the ability to predict the nature, likelihood, and potential severity of weather impacts, uncertainty remains. In addition, although many hazards have features that create unique decision-making environments, the potential to apply lessons learned from one hazard context to another is considerable.

In early 2024, the National Academies of Sciences, Engineering, and Medicine (National Academies) convened an ad-hoc committee to plan a workshop on risk communication around tropical cyclones. Sponsored by the U.S. Department of Commerce, the National Aeronautics and Space Administration, and the National Science Foundation, this workshop occurred on February 5 and 6, 2024, with participants attending virtually or in person in Washington, D.C. The workshop goal was to identify opportunities and challenges to communicating about extreme tropical cyclones as well as lessons that can be drawn from community engagement and communication concerning other extreme events (see Box 1 for the full Statement of Task). Toward this goal, the committee designed the workshop to include speakers who offer a wide range of perspectives from across decision-making and communication processes and to consider the information needs, capabilities, and motivations of different decision-makers and decision-making audiences for risk communication (e.g., government, industrial, public).

Over the course of the workshop, the committee heard from academic researchers in the social and bio-physical sciences, public policy, and other fields; public officials at the federal, state, and local levels; and representatives from the private sector. Their comments covered not only tropical cyclones but also other hazards such as hurricanes, flooding, earthquakes, and extreme heat. Discussants addressed various facets of risk communication, including the importance and difficulty of clearly communicating uncertainty to the general public; preparedness as a critical component of an effective response; and new or in-development risk communication products and technologies. They discussed localization of information across audiences and communities and the need for deeper understanding of how messages are created, distributed, and perceived. They also identified gaps in research and practice, such as a lack of real-time research opportunities and insufficient attention to the needs of specific populations, to inform future research agendas that are of demonstrated and critical interest to U.S. federal agencies, stakeholders, and decision-makers at all levels in extreme weather–prone regions of the United States and elsewhere.

The creation, continuation, and fortification of partnerships and the importance of collaborative work—particularly among decision-makers across jurisdictions, sectors, and population groups—emerged from the discussions as a common theme. The workshop served as a space for participants to not only discuss these important topics, but also make new connections or deepen previous collaborations. As Ann Bostrom (committee chair), Weyerhaeuser endowed Professor in Environmental Policy, University of Washington, noted in her closing remarks that the workshop itself provided connections on which to build.

ORGANIZATION OF THE WORKSHOP PROCEEDINGS

The two days of the workshop are described in these proceedings chronologically, with day 1 covered in Chapters 2, 3, and 4, a recap of the first day summarized in Chapter 5, and day 2 covered in Chapters 6, 7, and 8. Chapter 2 summarizes the first session, with panels on research and forecaster perspectives around risk communication pertaining to atypical tropical cyclones. The goal of this session was to gain an understanding of the unique challenges, opportunities for innovation, and lessons learned in communicating evolving tropical cyclone threats using lessons learned from the atypical tropical cyclones Henri (2021) and Hilary (2023). Chapter 3 summarizes sessions two and three, which are framed by a keynote speech on definitions and classifications and feature panels (session two) and report-outs from breakout discussions (session three) on lessons arising from multi-hazard events and other hazards that might be applied to the tropical cyclone context. Chapter 3 also contains a high-level summary of sessions one and two. Chapter 4

BOX 1-1
Statement of Task

An ad hoc committee will plan a workshop to bring together experts to explore challenges and learning opportunities around actionable and understandable risk communication with decision-makers for extreme weather events. In particular, the workshop may consider the information needs, capabilities, and motivations of different decision-making audiences for risk communication (government, industrial, public) in the service of protecting lives, property and livelihoods. Discussions will include issues of justice, equity and inclusion in risk communication and community engagement both with and for vulnerable and underserved communities.

Workshop discussion will consider the following topics:

1. Explore the current understanding of effective communication practices and features to convey to decision-makers uncertainty/probabilistic information about risks associated with discrete, extant extreme weather events. Discussions may include barriers faced by decision-makers in implementing uncertainty/probabilistic information, benefits and challenges with existing Impact-Based Decision Support Services (IDSS), and lessons learned in the light of recent events.

2. Examine risk communication and decision-making challenges posed by extreme weather events that are unprecedented in nature or scale for the affected locations. Discuss what communication practices and features are most effective for addressing these challenges, which may include accounting for historical precedence, diverse populations, and the impacts of climate change on the nature, behavior and frequency of extreme weather events as well as the potential for compounding or cascading events.

3. Explore opportunities for learning from synergies, successes and challenges across multiple hazards and decision-making contexts and applying them to the hurricane context. Discussions may include hazard or event types with different lead times, different motivations (or success criteria) among decision-makers, vulnerable communities or livelihood sectors with different characteristics, outcomes of communication that are considered both "successful" and "unsuccessful," and factors and strategies that contribute to successful community engagement and co-production of risk-reduction strategies.

---

summarizes session four, which was designed to inform understanding of various risk communication needs and sources across scales (e.g., county, municipal, faith-based organizations, local emergency management) and communities, spanning the household, local, and state levels, and the challenges that arise across population segments with differing experiences. Chapter 4 also provides a high-level summary of sessions three and four. Chapter 5 summarizes day 1. Chapter 6 summarizes session five, which featured panels on public- and private-sector innovations and a demonstration of new messaging technologies. Session 5 aimed to (1) examine current and emerging methodologies for communicating risk/uncertainty information in the public arena and (2) identify how new risk communication technologies and approaches are being evaluated for their effectiveness in communicating risk and motivating behavior in the public arena. Chapter 7 summarizes session six, which offered a keynote speech on the use of jargon, technical, and plain language and panels on communicating uncertainty and issues about access and functional needs in the context of risk communication. Session 6 aimed to highlight unmet needs in communities at risk from tropical cyclones and potential solutions to meet those needs in the context of communication. Chapter 7 concludes with a high-level summary of sessions five and six. Chapter 8 provides a workshop retrospective and final remarks. Appendixes A and B provide the workshop agenda and biographies of the planning committee members, respectively.

This proceedings summarizes workshop presentations and discussions and has been prepared by the workshop rapporteurs as a factual summary of what occurred at the workshop. The views contained in the Proceedings are those of individual workshop participants. The planning committee's role was limited to planning and convening the workshop. The views contained in the proceedings are those of individual workshop participants and do not represent consensus views or recommendations of the National Academies or represent the views of all workshop participants, or the study committee.

Chapter 2

Communicating Risks of Atypical Tropical Cyclones: Lessons from Henri and Hilary

Andrea Schumacher (committee member), Project Scientist, National Center for Atmospheric Research (NCAR), opened the first session of the workshop, which featured panels about lessons learned from the atypical tropical cyclones, Hurricane Henri (2021) and Hurricane Hilary (2023), from the perspectives of forecasters and researchers. Schumacher noted that this session aimed to inform "understanding of the unique challenges, opportunities for innovation, and lessons learned in communicating evolving tropical cyclone threats."

FORECASTER PERSPECTIVES ON RISK COMMUNICATION

Alex Lamers, Warning Coordination Meteorologist, Weather Prediction Center, National Oceanic and Atmospheric Administration (NOAA); Rose Schoenfeld, Meteorologist, National Weather Service Weather Forecast Office, Los Angeles; and Robbie Berg, Warning Coordination Meteorologist, National Hurricane Center (NHC), discussed risk communication from a forecaster's perspective. Following brief introductions, the panel took the form of a structured discussion, which was guided first by questions posed by the moderator, Schumacher, and then an extended question-and-answer session.

Schumacher first asked about challenges to communicating risks associated with rare and atypical tropical cyclone events. Schoenfeld noted that the language commonly associated with tropical cyclones does not apply in Los Angeles, California, because such events are rare and, when they do occur, their impacts differ from those experienced in more typical locations. Lamers observed that another challenge lies in understanding how meteorological information will "translate to impacts" in areas that do not normally experience such events, such as the rains associated with Hurricane Hilary that affected Death Valley.[1] Increased frequency of

---

[1] More information about the flooding impacts of Hurricane Hillary in Death Valley, California, is available at https://www.nps.gov/deva/learn/nature/hilary.htm.

extreme events is another challenge, she explained. Berg described two challenges that stand in tension with one another: the problem of "noise" (i.e., many messages from multiple sources such as social media and traditional media) and how noise can overwhelm people and overfocus the message on information less relevant to staying safe (e.g., landfall time, possibility of breaking records) while obscuring messaging from the NHC and other institutions about the risks that such storms pose. Lamers added that, although media often focus on factors such as atmospheric pressure and windspeed, the deadliest hazard in the United States in the past decade has been rainfall-induced flooding. Lamers and Schoenfeld both spoke about the challenge of relaying the hazards of intense rain events when the public might perceive these events to be just "bad weather."

Schumacher's second question to the group introduced a theme that surfaced repeatedly throughout the workshop: the importance of strong relationships between various partners. In their responses to Schumacher's question of how decision makers, including emergency managers (EMs), responded to warnings from the various institutions represented, each panelist emphasized the benefits of having strong long-term, working relationships between their national institution and local EMs and other local decision-makers. Berg noted that as weather changes, relationships between the NHC and certain locations, particularly concerning unusual types of events, may need to strengthen: "We don't see tropical cyclones hitting Southern California that frequently, so we don't have that pre-established relationship with many of those emergency managers on tropical cyclones themselves." Schoenfeld shared the hope that partnerships around "more routine" hazards would extend in the case of unprecedented events. Lamers noted that this extension seemed to be successful in the case of Hurricane Hilary, which was rare and costly but resulted in no casualties. He added that plans for targeted evacuations in New York City, developed after Hurricanes Henri and Ida, and for evacuations targeting vulnerable populations during Hurricane Hilary in California, revealed that decision-makers have a high level of confidence in the rainfall forecast and other information provided at the national level. Echoing another common theme—improving communication through better understanding of specific recipients' needs and wants for this information —Lamers said, "I think understanding a little bit more of how they are making these decisions . . . could really help inform us as we try to improve these forecasts over time." Berg highlighted the importance of local specificity: even though a region might be frequently exposed to a certain type of event, a particular locality within that region may not have direct experience and, therefore, not the same level of familiarity. This specificity makes a difference in how populations understand risk and when and how they take action.

The discussion then turned to novel or innovative communication approaches during rare events. Schoenfeld shared how, in anticipation of Hilary, her office "pushed to get the timeline for products issuance accelerated," knowing that a tropical storm watch or warning had never been issued for the region. Berg also mentioned that accelerated timing was part of the NHC's messaging around

Hurricane Hilary, and Lamers described how the Excessive Rainfall Outlook tool can help visualize the extent of risk and influence (i.e., accelerate) the timing of messaging around unprecedented events.[2] Berg highlighted two other novel approaches at the NHC. The first is "Key Messages," which are designed and used to highlight essential points about hazards and forecast uncertainty for select tropical cyclones.[3] Key Messages enable consistent messaging through local channels. The second is livestreaming on social media and other platforms (e.g., Facebook, NHC YouTube channel).

Schumacher then asked the panel to reflect on what they might have done differently during recent events. Lamers emphasized the importance of talking to partners and taking their feedback seriously. "Rainfall rates" in urban areas, as a specific aspect of vulnerability and exposure, was of profound importance after Hurricanes Henri and Ida, he noted.[4] Therefore, the National Weather Service is developing an Urban Rain Rate Dashboard, a tool that will involve hydrologic modeling of a city to help decision-makers in large cities understand when rainfall rates are spiking in their areas. Lamers noted that "several dozen cities" will be mapped and modeled initially, with perhaps more to follow.

Schoenfeld described challenges to communicating uncertainty and associated concerns about losing credibility with the public. She noted that in response to feedback from the public, her office now more actively shapes and narrows the focus of warnings as the event unfolds. Berg returned to the concept of noise in messaging and the obstacles to helping people focus on the most relevant hazard in a changing situation with multiple hazards; he described, specifically, how people might hear "hurricane" and miss the danger of hazards such as rainfall, tornadoes, and rip currents, among others.

DISCUSSION

Questions posed by audience members spanned a range of topics. Sara McBride, Research Social Scientist, U.S. Geological Survey (USGS), asked about the usefulness of ARKStorm in risk communication, a worse-case hypothetical storm

---

[2] The Excessive Rainfall Outlook tool is "a graphical map issued by the Weather Prediction Center (WPC) that forecasts the probability that rainfall will exceed flash flood guidance (FFG) within 25 miles (40 kilometers) of a point across the contiguous United States (CONUS)" (National Weather Service, 2023). More information about the tool is available at https://www.weather.gov/media/notification/PDDs/PDD_ERO_Days_4_5_T2O.pdf.

[3] The Key Messages graphic on the NHC website and NHC social media pages "includes the text of the Key Messages and relevant tropical cyclone graphics, which can include the cone graphic, the 34-kt cumulative wind speed probability graphic, or a rainfall forecast graphic provided by the Weather Prediction Center." More information about Key Messages and their associated graphics in the context of Hurricane Ian are available at https://www.nhc.noaa.gov/aboutnhcgraphics.shtml#KEYMESS.

[4] Rainfall rates "measure the intensity of rain within a certain period of time" (Weather Nation, 2021). More information about rainfall rates is available at https://www.weathernationtv.com/news/understanding-rainfall-rates.

scenario model that is used as a learning tool to provide EMs, the public, and other groups with an assessment of what is historically possible.[5] Schoenfeld and Lamers both noted that messaging the true worst-case scenario is not always ideal in the realm of public communication, because it can distract and obscure information relevant to particular hazards or locations. However, they noted that worst-case scenario exercises, such as ARKStorm, can be used to develop preparedness protocols, helping EMs and officials to envision what atypical events might look like in their area or to stress-test emergency response systems. Karen Florini, Vice President for Strategic Impact, Climate Central, asked about the way that rapid intensification events affect warnings, such as a tropical depression rapidly and unexpectedly intensifying into a high-category (e.g., 3-5) hurricane. Berg acknowledged that such events are occurring more frequently but that statistics indicate that forecasting of such events is improving.

Questions from the audience then turned to best practices for risk communication in partnership with other countries and continents. The strength and importance of partnerships were again highlighted, as exemplified by the NHC's coordination with countries in other regions (e.g., the Caribbean). However, Berg stressed that, although the United States can and does make recommendations to other countries, each individual country is responsible for deciding whether and how it issues warnings and other messages. He added that the Caribbean and Central America are the only regions in the world where such coordination and partnership occur between the country responsible for tropical cyclone forecasts (in this case the United States) and other countries in that region. National hurricane centers exist in Japan, Australia, Fiji, and other places, "but they don't have the framework that we have in our part of the world where they are coordinating with other countries around them quite as closely as we are," and those centers perhaps should consider strengthening international partnerships "in order to get those risk messages out to other countries that aren't actually making the forecasts themselves," acknowledged Lamers. In addition, Lamers noted as a best practice that the NHC in Miami includes all hazards in its official public advisory products, which garner widespread attention. Official advisories from the NHC in Miami include hazards stemming from the potential for rainfall and flooding, storm surge, tornadoes, and wind, in addition to the more standard forecast track and intensity. Such an approach could be adopted by other regional warning centers around the world.

In response to the final question, the panel described how strategies change as lead-time changes.[6] Schoenfeld emphasized the importance of finding ways

---

[5] ARKStorm was developed by NOAA, USGS, Scripps Institute of Oceanography, State of California, California Geological Survey, University of Colorado, Federal Emergency Management Agency, NCAR, California Department of Water Resources, California Emergency Management Agency, and other organizations. It addresses "massive U.S. West Coast storms analogous to those that devastated California in 1861-62 and with magnitudes projected to become more frequent and intense as a result of climate change." More information is available at: https://www.usgs.gov/programs/science-application-for-risk-reduction/science/arkstorm-scenario.

[6] Lead-time is "the difference in time between the onset of an observed event, and the issuance of a forecast that is associated with the observed event" (Lough et al., 2008, p. 5). More information is available at https://gsl.noaa.gov/fiqas/publications/articles/TAF_Leadtime_Metric_FULL_Description.pdf.

to communicate uncertainty, both to the public and in conversation with partners, while Lamers noted that decisions about lead-times should account for the rhythms of daily life—for example, timing a message so that people receive it before embarking on weekend travel.

RESEARCH PERSPECTIVES ON RISK COMMUNICATION

The second panel of the session complemented the first, with researchers sharing their perspectives on the same topic of risk communication around atypical storms. Ann Bostrom moderated this panel.

Roxane Cohen Silver, Distinguished Professor of Psychological Science, University of California, Irvine, spoke first about her work to determine what makes research "ideal" from a methodological point of view, and then on her findings from research using new methodologies on risk perception around evacuation zones. Silver and colleagues conducted a literature review of "decades of research on how people evacuate from natural disasters, whether or not they do, and the characteristics of this body of research" (see Thompson et al., 2017). The research team noted that the high number of conflicting results was due to methodological limitations and deemed "ideal research" to be research that identifies at-risk population samples before an event, takes a longitudinal approach with immediate and then repeated post-event assessments, and uses representative samples of the subject population.

The research team then designed a study based on these ideal characteristics to explore the psychological impact of Hurricane Irma (Garfin et al 2022). The team collected data at two points: (1) about 60 hours before landfall, from about 1,600 people, in a representative sample and (2) 1 month later, in a follow-up assessment, with a representative sample of 1,500 people. Findings showed that about 40 percent of participants did not correctly identify their evacuation zone status, about 6 percent were unaware of evacuation orders, and about 17 percent evacuated unnecessarily (Ibid.). The strongest predictor of whether people evacuated is whether they had evacuated for a previous storm, independent of whether they were in an evacuation zone. Similarly, pre-hurricane risk perceptions were a strong predictor of whether a person evacuated. However, Silver asked, "What predicts the pre-hurricane risk perceptions?" Media exposure and clarity of messaging—particularly visual messaging—were the strongest predictor, said Silver. She highlighted a mixed public message wherein the text instructed readers to evacuate only if they were in an evacuation zone, but the image showed a cone across the entire state.

Silver concluded her remarks with three messages: (1) it is important to make changes to communication practices as well as research; (2) risk perceptions and perceptions of evacuation zone status appear to shift over time, and perceptions of zone status may not be accurate; and (3) partnering with media during the event to craft and update evacuation messages that accurately reflect changing risks is critically important, and debriefing after the storms with both the media and the general public can increase understanding and preparation in anticipation of future storms.

Julie Demuth, Project Scientist III, NCAR, presented on her research on how risk perception, risk information, and protective action evolve as the event itself evolves. Like Silver, Demuth investigated limitations of current methodologies, developed ways to address these limitations, and applied them in her own work. Her remarks focused on the findings of a longitudinal study of risk perceptions during Hurricanes Henri, Laura, Marco, and Ian, comparing the people's responses to Henri—an atypical storm—to people's responses to the other three, more typical events (see Demuth et al., 2023).

The study collected data across four waves of assessment—three during the predictive phase of the storm and one after the storm, Demuth explained. With these data, the research team tracked the frequency with which people received information, and from what source, including environmental cues; the relative importance of different types of information; risk perception, based on respondents' estimation of the likelihood that their area would be affected by the different hazards; and whether respondents took protective action such as buying supplies, boarding up windows, or evacuating. Demuth highlighted the importance of asking about the effects of the different hazards associated with a single event.

Findings revealed that people received information from the NHC less frequently during Hurricane Henri compared to the other hurricanes. Demuth explained that more people reported "looking outside" to gather information during Hurricane Henri compared to other events, with a large increase between waves two and three. Regarding the relative importance of different types of information, among Hurricane Henri respondents, the cone of uncertainty and wind speed became increasingly important between waves one and two.

Data for risk perceptions disaggregated by hazard showed that Hurricane Henri respondents reported a large increase in concern from wave one to two, with wind as the largest area of concern. Negative impacts of particular concern included power outages and road closures. For Hurricane Henri, compared with other storms, respondents were less likely to believe they would experience emotional impacts or financial losses. Finally, regarding the question about protective actions taken, Demuth reported that, overall, the data from Hurricane Henri are comparable to those from other storms in some categories (e.g., following the forecast, moving things, and doing other home preparation) but protective actions were lower in others (e.g., getting supplies, gassing up their vehicle, boarding up, evacuating).

Demuth echoed Silver's comment on the importance of this particular methodology to help identify ways to improve risk communication: "This kind of longitudinal, perishable data that we're collecting during a hurricane event, as it's actually threatening, is really essential to understand the dynamic processes that people are going through."

During the third and final presentation of the panel, Emma S. Spiro, Associate Professor, University of Washington Information School, discussed research on the role of social media during a crisis event—including natural hazards, but also civil or political unrest, domestic terrorism, or breaking news such as for elections.

Recalling the reference to noise during an earlier panel, Spiro explained that this noise is, in a sense, her primary research topic: the flow of information within social media environments and what people do with the information they encounter in these environments. In the midst of a crisis event, people "come together to try and make sense of what is going on around them" and to find information that might help them make decisions—a "collective sense-making process," as she called it. Commonly, rumors emerge in these situations, Spiro explained, defining rumor as "information unverified at the time it's being talked about." Rumors can be useful when they bring people together, reduce anxiety, and spur people to take protective action. Challenges arise, however, when the collective sense-making process is strategically manipulated through mis- or dis-information.

Spiro described findings from an ongoing study that she co-leads at the University of Washington that analyzes millions of messages sent on select social media platforms for message content, format, design, and how people engage with information in order to identify and better understand the "rules" that govern information flow. These rules include levels of uncertainty and trust of the social media platform, emotional valence, the actual words, and larger narratives and worldviews within the different communities affected by the event. Spiro stated that the study has helped to identify "the features of different information narratives of rumors that contribute to their virality in online systems" and has shown that people engage with these systems and spread information largely because they want to help each other.

The social media information environment poses widespread challenges for researchers, EMs, and the public, Spiro noted. The system is participatory by nature—people have to participate in order for information to be spread online—and therefore vulnerable to manipulation by bad actors who use participants to spread mis- or disinformation. The emergence of artificial intelligence (AI) and deepfake images increases the complexity of information and vulnerability of the system: "We have challenges around not only the increased quantity and quality of information and a very low barrier to doing that [i.e., creating a false image] but also increased personalization of information, persuasion of information, and other open pathways for these tools to involuntarily produce false or misleading information." Spiro also noted that these manipulations are not entirely new, referencing an image of a shark swimming down a highway that tends to pop up on social media during a tropical storm or hurricane.

Researchers face challenges as well. Although critical to understanding information flow during a crisis, data on social media use are becoming more difficult to collect as social media platforms increasingly limit access options for researchers, Spiro explained. Fragmentation of the social media environment, with multiple platforms that users move across, also complicates observation of behavior and information flow. Similarly, "multimodal information" that combines audio, video, and text is becoming more common and presents new challenges. Spiro concluded by emphasizing that trust in information and institutions is down, which hampers research and undermines trust in other people more generally.

Bostrom, as moderator, posed the opening question for the group discussion session. She asked how the experiences of the broadcasters and EMs described during the first panel might connect with their own research agendas. Demuth highlighted the need to establish a common understanding of the term "risk communication," adding that it could be thought of as a frame that connects different research areas, as well as the research space to forecasting work. This broad concept encompasses not only the messages to the public, she remarked, but also the tools available to forecasters, the challenges they face in predicting and timing, and the ways that researchers might help deepen forecasters' understanding of information as they make decisions about risk and generate new messages for partners and the public.

In her response, Silver emphasized the need for partnerships, including among EMs and people in the weather community who "help us figure out what kinds of questions we should ask." Shaping and improving research questions, methodologies, and communication strategies by intentionally and carefully attending to the particularities, preferences, and feedback of the community on the receiving end of research and messaging is important. Silver cited an example from her current research in Lake County, California, a community at a high risk of wildfires. Although 60 percent of the community has suffered wildfire damage since 2015, 60 percent of respondents in the research sample said they did not know their evacuation zone, while 40 percent of those who said they did were mistaken. Her research team is sharing these data with EMs and is exploring other ways to support risk awareness.

Spiro noted the important but challenging work of lending expertise in real time, working with journalists, for example, to ensure high-quality information as events unfold. She noted that such partnerships push researchers in "thinking about how research might have to operate in a very rapidly changing environment."

Returning to the topic of evacuation zone awareness, Richard Allen, Director, Berkeley Seismology Lab, University of California, Berkeley, asked whether it was possible to develop targeted messages to the cell phones of people in affected evacuation zones, similar to early warning earthquake messages. Silver noted that EMs working to improve evacuation zone awareness have noted information retention as an issue. Another challenge is that the evacuation zone is represented by a complex set of numbers and letters that are unfamiliar to the public. Christina Finch, an audience member, added that evacuation zones are developed at local levels. Therefore, there is neither a national standard for developing the zones nor a "consistently available resource to find all zones." She wondered whether an individual's understanding of their evacuation zone and ability to remember this information are influenced by how zones are defined and the types of information and messaging (e.g., maps, apps) that are available in particular jurisdictions. Silver responded that a website was designed during Hurricane Irma to show the zone for any specific address entered, but she could not get useful information from it.

Her research showed that people do not know their zone, but there is a lot of information about how and why this issue exists, and how it can be addressed. Demuth added that her research similarly shows people are incorrect in their perceptions about evacuation zones. However, she also urged the expansion of knowledge around evacuation zones, including a more nuanced understanding of the multiple hazards in a complex weather event in the context of evacuation. "How do we map these impacts with some of these extreme events or these atypical events when we might not have evacuation zones that help us understand people's actual risk?"

The last question looked to the future, with an audience member asking the panelists to describe the types of new research they might employ to improve risk communication. Spiro noted that one important step is to explore the most effective ways to share information across platforms, while developing better ways to observe how information moves across platforms. She also mentioned the need to better understand how information from official and unofficial sources can be used and the potential benefit of leveraging AI tools. Silver commented on the difficulty and expense of conducting pre-disaster, real-time studies, with the pressures of timing making pre-disaster funding critical, along with the need for quick assessment and approval of projects from the institution's review board. Demuth said she hopes for more social science research of this kind and echoed Spiro in calling for further development of real-time transfer of knowledge from researchers to the "operational community" in the midst of an event.

Chapter 3

Risk Communication in Multi-Hazard Environments: Challenges and Learning Opportunities from Compounding Hazards and Cascading Impacts

Marshall Shepherd (committee member), Georgia Athletic Association Distinguished Professor of Geography and Atmospheric Sciences, University of Georgia, introduced the second session about lessons that can be learned from compounding hazards and cascading impacts. He outlined the committee's four goals for the session: (1) gain a better understanding of how compounding and cascading hazards are defined; (2) explore the effects of recent events such as Hurricanes Harvey, Laura, and Michael on the vulnerability of infrastructure and at-risk communities; (3) characterize unique challenges associated with risk communication around these kinds of events; and (4) explore opportunities for advancing risk communication particularly around compounding hazards and cascading impacts. In a keynote talk, Jen Henderson, Assistant Professor of Geography, Texas Tech University, discussed the complicated and consequential work of classifying or defining events, particularly in the context of increasingly more frequent extreme or atypical storms. A panel followed the keynote talk and included, Jason Senkbeil, Director of Undergraduate Studies & Professor, Department of Geography, University of Alabama; Rebecca Moulton, Meteorologist, Federal Emergency Management Agency (FEMA); Jeff Lindner, Meteorologist, Harris County Flood Control District; and Jessica Schauer, Tropical Cyclone Program Leader, National Weather Service. The panel addressed questions about risk communication around compounding hazards and cascading impacts.

KEYNOTE SPEECH
MULTI-HAZARD EXTREMES: DEFINITIONS, CLASSIFICATIONS, AND CONSEQUENCES

"Classifications and the way we talk about risk are not benign; [they have] consequences that are material," Henderson declared, setting the stage for her keynote speech, which focused on the intersection of extreme storms and classification practices in the context of risk perception. Referencing work on organization and categorization by Bowker and Starr (2000), Henderson contended that categories and definitions have high moral, ethical, and social stakes. Questions about how to classify extreme weather events have become more urgent as such events have become more frequent. Extremes are shaping a "new normal," she noted, where increasingly frequent atypical and unprecedented events put pressure on traditional classification systems. "The traditional classification of natural, human-made, and hybrid disasters seems to be an insufficient tool in the face of the high complexity of the present-day world," she explained. In thinking about the stakes of this issues, Henderson looked at both sides—first extremes and then classification itself.

Although extreme storms are generally understood in terms of "billion-dollar hazards," a category designated by the National Oceanic and Atmospheric Administration (NOAA) in 1980, this one-size-fits-all designation can exclude other meaningful ways to think about extremes, she noted. Half- or quarter-billion-dollar hazards might be considered extreme depending on the population affected and the frequency of occurrence. Storms might also be classified using intensity, death toll, and other metrics. The increasing frequency of extreme events—however they are measured—demands rethinking collective and individual understandings of the "billion-dollar hazards" category, which in turn would shift ethical frameworks and practical responses to these events, Henderson argued. She highlighted pollution as a helpful analog to extreme storms—something rare that became normal—and noted that the answers to such questions would affect risk communication itself and the types of research conducted.

Classification itself becomes an issue as hazards evolve and this "new normal" emerges, Henderson explained, and one with serious consequences. She highlighted the need for new categories as atypical storms occur more frequently and as climate change introduces new levels of uncertainty. Existing categories are also being challenged by new forms of interconnectedness; increased and different social, technical, and physical connections can create "more magnified emergencies [that] co-occur in time and space," she noted.

The terms "cascading" and "compounding" present particularly thorny classification challenges. One challenge lies in definitions: the term "cascading hazard" can denote both the unexpected secondary events that flow from an originating event and an event that emerges from "a series of connected errors and failures that create the conditions for a greater malfunction and more devastating consequence." Henderson explained that the literature on compounding hazards is similar to that of cascading impacts in that it tends to repeat a general understanding of the concept as "a combination of multiple drivers and/or hazards that contribute to societal or environmental risk." Like extremes, compounding hazards and cascading impacts are occurring more frequently to the point of being "ubiquitous," and, thus, put pressure on traditional categorization. For example, typologies and models exist for classifying and thinking about compounding hazards and cascading impacts; however, she explained, these categories speak primarily to physical infrastructure and drivers, and often do not account for the impact on social and technical factors or their role as driving forces.

Henderson highlighted several conceptual models for thinking more expansively about cascades and compounds, such as how extreme events can produce multiple hazards that in turn result in cascades that trigger emergencies and co-occurring compounds (e.g., technical, physical, social) (Figure 3.1). The sociologist Susan Cutter, for example, developed the idea of "social cascades" to refer to "the social, cultural, economic, and political effects of consecutive disasters within either close temporal or close spatial proximity" (Cutter, 2018, p. 23), arguing that they should be included in research alongside physical drivers. One source for such models are media that take up topics not usually discussed in the research world; in this vein, Henderson flagged journalistic material, including "Floodlines," a podcast produced by The Atlantic that traces multiple complex, intersecting elements around Hurricane Katrina, as well as work by Dave Eggers, and Kai Erikson and Lori Peek.[1]

According to Henderson, the consequential way that classification impacts communication can be clearly seen during multiple simultaneous events, where warnings and responses for each hazard can contradict one another. For example, in the case of TORFF events (Tornadoes and Flash Floods that happen simultaneously), warnings that deliver conflicting information to people about what they should do often overlap (i.e., come within 30 minutes of one another). Henderson noted that more than 400 such overlaps occur each year in the United States (see Nielsen et al., 2015). Tropical cyclones, which often involve multiple overlapping

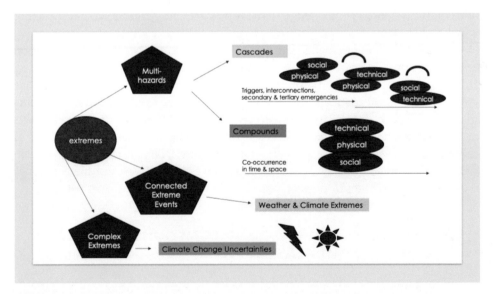

FIGURE 3.1 A visual representation of how the literature has framed different types of extreme weather and climate hazards.
SOURCE: Presentation by Jen Henderson on February 5, 2024; created by Jen Henderson.

---

[1] More information about the "Floodlines" podcast is available at https://www.theatlantic.com/podcasts/floodlines/.

wind and water threats (e.g., synoptic winds as well as tornadoes; storm surge as well as flash flooding), have an even greater potential for overlap. Further complicating matters, overlaps can change as timing and scale of the multiple hazards evolve, she explained. Cascading hazards present similar communication challenges.

Henderson elucidated four main communication challenges around multi-hazard events involving tropical cyclones. First, as in the case of TORFFs, instructions may conflict (e.g., sheltering is different in the case of a tornado or flood). Second, the increased complexity that comes with dealing with multiple hazards and how messaging and prediction often involve multiple agencies, different spatial and temporal scales, and threats that move and evolve. A third challenge, which stems from the previous two, is that the public may have a higher risk perception for one hazard than the other, regardless of which hazard is most threatening in a given storm (see Henderson et al., 2020). News coverage and policies driving messaging from various agencies about risks for flood and tornado hazards can "unintentionally magnify one hazard" over the other instead of helping people to prioritize which hazard to understand as the most threatening. Finally, communications experts are experiencing increased stress and strain due to the more frequent and intensifying extreme events.

Classification decisions have a ripple effect with social and ethical dimensions, Henderson noted. They can, for example, determine who has an advantage and who suffers; which regions benefit more or less; and who keeps or loses jobs. Categorization decisions also shape how knowledge is built and shared. Classification underpins disciplinary knowledge that evolves around different hazards, each of which has its own "epistemic culture of risk" as well as research agendas, funding mechanisms, labs, technologies, and policies Therefore, the nature of classification systems can inhibit interdisciplinary and convergent thinking, Henderson argued. However, a lot of opportunity for expansive, cross-disciplinary thinking lies in the work around social cascades and compounds. Henderson concluded by stressing the importance of "integrative, convergent work" that looks at cascading impacts at multiple temporal and spatial scales; attends to unanticipated consequences of compounding events; and—especially—takes a holistic, flexible approach that classifications do not always allow. Such research would also "re-examine the things that classification hides or doesn't make visible for us," Henderson noted.

RISK COMMUNICATION AROUND COMPOUNDING HAZARDS AND CASCADING IMPACTS

In the panel that followed Henderson's keynote speech, speakers took up different angles of a theme that resonated throughout the session as a whole: the complexity of risk communication about events with cascading and compounding hazards, especially as extreme storms occur more frequently and have unanticipated impacts.

Senkbeil began by elaborating on two themes related to cascading and compounding hazards: (1) an increase in hurricanes at the tail of distribution in recent extreme events, for example, the number of hurricanes that have rapidly intensified within 400 kilometers of the coastline has tripled since 1980 (Li et al., 2023) and (2) the potential for social science to reveal how people understood previous hurricanes and to deepen understanding of how previous experiences impact their current perception of risk. Although every storm is unique, in multi-hazard storms, either water or wind can pose a greater threat and distinguishing between the two can help produce more effective communication, Senkbeil explained. For example, windstorms are fast-moving, with less rain and a lower surge volume but a higher peak surge, while water storms tend to be slow-moving, bringing more rain and a larger storm surge volume, he noted. Communicating "the greatest hazard of concern," as Senkbeil called it, can help people understand risk and take appropriate action.

Another major challenge to risk communication lies in enabling the public to understand the risks posed by events that are more extreme than those previously experienced, Senkbeil noted. People often have a "benchmark storm"—a previous event against which they measure risk in the present; however, as compounding, cascading hazards increase, past experiences are not always comparable to what might happen in the "new normal," and benchmarking can lead people to underestimate the severity of the present threat. However, Senkbeil suggested, this thinking could be leveraged by emphasizing the potential for the current event to exceed the "ceiling" of the benchmark event. It can be especially difficult to get the public to truly understand the impacts and risks of water storms, he said.

Messaging and predictability are also challenges during what Senkbeil termed grey swan events—storms that are predictable but very unlikely ("in the skinny tails of the distribution"), which often intensity rapidly within the 24-hour period before landfall. He provided two examples from Hurricane Michael, when rapid intensification meant that people experienced much stronger winds than expected and experienced impacts in new places.

Senkbeil also discussed several avenues to improve risk communications. First, communications that emphasize the hazard of greatest concern, especially distinguishing between wind and water storms, is extremely helpful. This approach could also enhance the public's understanding of the risks posed by cascading threats that accompany each type of storm and each hazard within a multi-hazard storm. Second, the new NHC cone of uncertainty, which shows more inland impacts, is a positive step toward clearer communication about the cascading impacts of storms. Finally, communications that compare predictions for a storm in progress with the values for previous, benchmarked storms to which people often refer can help to set accurate expectations and understandings about risk.

Moulton spoke about the importance of listening to different audiences and partners to effective communication, as well as how her own experiences shaped her understanding of such work. Through her work supporting local emergency managers (EMs) and evacuation messaging, she has realized the central importance of listening to the audience, understanding their questions and problems, and talking directly with them whenever possible. Moulton shared that attentive listening over the years prompted her to reframe her own role. Rather than focusing solely on the presentation of accurate data, she embraced an approach grounded in seeking to understand the needs of her various audiences. Now, her guiding question is "How can I see things from their perspective . . . and let their needs inform all stages of our process, from planning to the briefings operationally and during the incident response?"

An approach grounded in listening and direct engagement is especially important, Moulton noted, because an enormous amount of information is freely available to everyone. Consequently, expertise in risk communication is not solely about the delivery of accurate information. Rather, one key to effective risk communication lies in figuring out what helps people understand information: what to pay attention to, what the context might be, and how to deal with uncertainty, for example. Probabilistic products are also appearing more and more, while deterministic products are waning. This change drives the shift from delivering information to helping people understand it, Moulton explained. She noted that this approach enables meteorologists to provide context as needed, "select the right information" from all the unknowns, and help individuals better navigate a large amount of uncertainty. It also gives space for the expertise of others: the EMs and other local officials, members of the public, and other partners who best know their situations and needs.

Navigating complex storm events with multiple hazards likely involves collaboration among individuals with specific areas of expertise, including meteorologists, other scientists, EMs, and local officials and community leaders, across multiple locations and agencies. Moulton noted the importance of listening to "subject matter experts" as part of this essential collaborative work. Along with the multiple perspectives that come with subject matter expertise, she added, this approach provides room for the human element of this work, where people are in stressful situations and often dealing with multiple events at once. Moulton's remarks captured what would emerge as a consistent theme throughout the workshop: the humanity of everyone involved in decision-making and communications during these events, including the public. Underneath all the layers of expertise, roles, and details of the specific situations, "we're all people," Moulton said. She concluded with an anecdote from her own life, recalling a moment when, while working for the White House Interagency Working Group on Extreme Heat, her own air conditioning broke. In that moment, she said, her perspective slipped from the outside expert to "one of the people."

Jeff Lindner spoke about risk communication particularly related to Hurricane Harvey. He, too, emphasized the complexity of risk messaging around a multi-hazard event, especially one that impacts a large geographical area. For Hurricane Harvey, this situation meant that the forecast for the mid-coast of Texas was a category 4 impact, with heavy winds and storm surge, but the forecast for the upper coast, including Houston, was for flooding from inland freshwater rainfall. Interpersonal communication across geographical regions can further complicate messaging about multiple hazards and appropriate actions to take. For example,

Lindner explained, people in the northern part of Harris County may hear that friends and family in coastal areas are being told to evacuate because of the storm surge, and wonder whether they too should evacuate, although their area is threatened not by a storm surge but by rainfall flooding (which usually does not trigger an evacuation in Houston or Harris County). He reported that in areas that do require evacuation, evacuation zones were mapped onto ZIP codes, so that most people knew their situation.

Linder shared two major challenges to risk communication: (1) helping people to understand the forecast information conveyed by various products and (2) helping people to understand the potential impacts of forecasted events. Although his comments focused on misunderstandings and mis-readings of messaging in relation to Hurricane Harvey, he noted that such challenges are not unique to that event and that lessons learned are applicable to many different instances and products. Lindner used a graphic produced by the National Weather Service (NWS) that predicted rainfall and catastrophic flooding in the Houston area during Hurricane Harvey to highlight a common misreading that underpins misperceptions of risk (Figure 3.2). People tend to believe that if they are not in the "bullseye" (i.e., the maximum amount of rainfall), then they will be fine: "I'm only going to get a foot of rain; the rainfall is spread over five days, so even 35 inches isn't so bad." Lindner stated that the rainfall rate is a salient issue with tropical systems and flooding. He noted that risk communications about rain for Hurricane Harvey may not have clarified the fact that rainfall would not occur at a steady rate over time, but in intense bursts that would result in "10, 15, 20 inches in a 12-hour period." Also, during Hurricane Harvey, members of the public focused on the fact that the hurricane itself would likely hit another area and did not understand that they might be impacted by far-reaching effects.

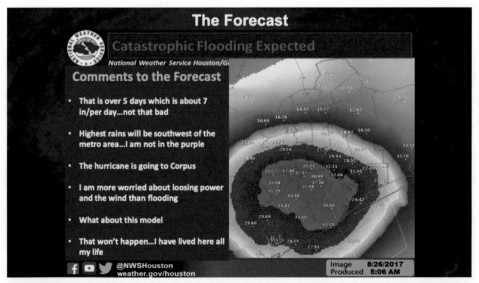

FIGURE 3.2 Forecast map of the Houston area during Hurricane Harvey with expected rainfall in inches and flooding indicated by color, alongside public comments about this forecast compiled by Lindner.
SOURCE: Presented by Jeff Lindner on February 5, 2024; created by National Weather Service.

Model forecasts—particularly those showing extreme scenarios—can also challenge clear communication and risk perception, Lindner noted. These models are often deterministic in nature and therefore differ from the official forecast, which foregrounds probabilistic information; and they are sometimes included in media broadcasts and reach the general public. Further, Lindner noted, such deterministic models can also influence decision-makers in the emergency management community, who would ideally rely on the official forecast.

Lindner echoed Senkbeil's comments on benchmarked storms: people understand risk of current/forecasted events through the lens of their experiences of prior events and assume that no storm will be as bad or worse as their benchmark storm. This assumption become especially problematic with multi-hazard storms because people do not anticipate extensive storm surge flooding, as evidenced, for example, in responses to Hurricanes Katrina, Ian, and Charley: "I just could never believe the water would get this high." Benchmarking contributes to a dangerous lack of understanding around risk in the case of rare or extreme events for which there is no historical context, Lindner said. Meteorologists can see "outlandish forecasts" (e.g., 50 inches of rain) that do not accord with previous experiences, and which people—whether the public or EMs—doubt or use valuable response time to verify. In these cases, different sources must be on the same page and distribute the same information to build confidence in the forecast.

A related challenge introduced by extreme, multi-hazard events is the pressure they can put on meteorologists to discern the line between forecast and impact, which can cross into areas that lie beyond their expertise, Lindner explained. "Our job is to forecast [and] to explain those forecasts, but . . . we don't have all of the knowledge for all of the impacts that can happen in a certain situation." It is not always clear whose job it is (e.g., forecasters, local officials, EMs) to know about and communicate about all possible impacts. Contradictory call-to-action statements, discussed by Henderson and Senkbeil, also confounded risk communication during Hurricane Harvey, Lindner noted. That storm affected a large geographical area, and thus products seemed to contradict one another because they addressed different impacts at different locations. Different hazards also affected the same area, leading to conflicting instructions: tornadoes prompted calls to retreat to the lowest floor of the house, while flash flood warnings had people going onto their roofs. In this situation, Lindner said, the Emergency Operations Center (EOC) and local NWS Office decided to keep the focus on flooding as the primary threat because the tornadoes were short lived and weak. Relaying the risk of a tornado often seems to take precedence over relaying the risk of flooding in people's minds, Lindner explained, so it was important to take steps to keep the flooding risk at the forefront. In summing up, Lindner emphasized the importance of discerning: "What is the primary threat right now?" Not every event is the same, so each requires discernment of the primary threat and then creation of appropriate messaging— "putting it out there, and keeping it out there, and making sure that hopefully what you're asking people to do matches what the highest threat is."

Jessica Schauer rounded out the panel with a discussion of how risk communication originates—that is, how different agencies and platforms coordinate and amplify key messages—and how to ensure consistency downstream. "Every storm is different," Schauer noted, a sentiment that resonated with her fellow panelists' emphasis on case-by-case decision-making and attention to local knowledge. These differences highlight the importance of close and nimble coordination around messaging as events unfold. At the NWS, a decision support services coordinator will work within the agency with forecast offices, river forecast centers, and the National Water Center, among others, to ensure that products include the NWS's key messages and are consistent. [2]

Schauer shared a series of standard graphics that are distributed whenever there is a tropical cyclone watch or warning within the Continental United States, Hawaii, or Puerto Rico/U.S. Virgin Islands. Built on probabilistic information, these graphics show "a reasonable worst-case scenario for each of the four hazards" (Figure 3.3) for a tropical cyclone. They are intended to be used by EMs, other partners, and the public to decide how best to prepare for the storm.

Using these graphics, Schauer illustrated three important aspects about the creation and coordination of risk communication, all of which served as touchstones throughout the workshop: (1) the desire for probabilistic rather than deterministic information—and the challenge around helping the public to accurately understand products that discuss risk in probabilistic ways, (2) the desire for localized information, and (3) the challenge of issuing potentially conflicting calls to action

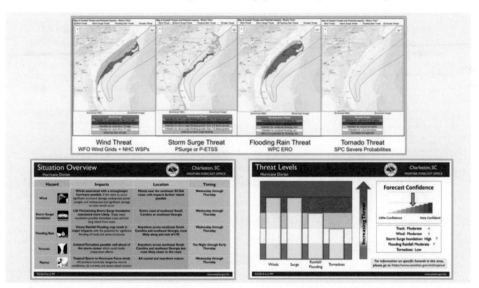

FIGURE 3.3 Hurricane Dorian (2019) threat graphics (top) that incorporate probabilistic information to provide a reasonable worst-case scenario for each of the four hazards (i.e., wind, storm surge, flooding, tornado), a situation overview (bottom left), and the four hazard threat levels relative to each other with a forecast confidence rating (bottom right).
SOURCE: Presented by Jessica Schauer on February 5, 2024; created by National Weather Service.

---

[2] More information about the National Water Center is available at https://water.noaa.gov/about/nwc.

for multiple simultaneous hazards (i.e., going to the lowest level of a house to prepare for a hurricane versus going to a higher level of a house to escape flooding). [3]

The public and relevant user groups (e.g., EMs) had requested more localized information and graphics such as those presented in Figure 3.4, Schauer explained. These graphics are issued by local weather forecast offices with clickable interfaces so that users can zoom in to local levels. They also provide information on which hazard poses the highest threat to a particular area. In the case of multiple co-located hazards, social science research could help discern how best to rectify conflicting calls to action. Information about direct and indirect fatalities in past events might be used to create more tailored messaging in the future, she added.

Comparing Hurricanes Laura, Michael, and Harvey, Schauer noted that, although all were multi-hazard and rapid intensifiers, each involved different hazards that posed threats not only during the active storm but also in its aftermath. For Harvey, flooding often prevented medical access in the wake of a tornado. For Michael, the large impact area complicated recovery after the storm. For Laura, widespread power outages meant that many people struggled to power their homes while those with generators risked carbon monoxide poisoning. Attending to post-event messaging is critical too: "The event isn't really over when the weather part of it is over."

The question-and-answer portion of the panel opened with a question from

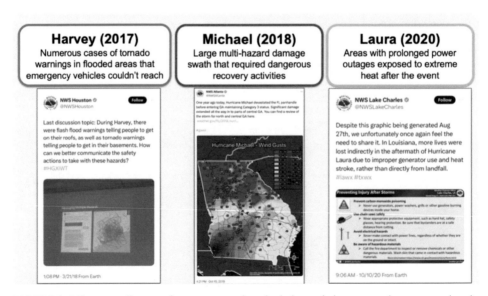

FIGURE 3.4 Three rapidly intensifying storms with multiple hazards (e.g., tornados, extreme heat). SOURCE: Presented by Jessica Schauer on February 5, 2024; created by National Weather Service.

---

[3] More information and graphics about hurricane threats and impacts is available at https://www.weather.gov/media/srh/tropical/HTI_Explanation.pdf.

Sunny Wescott, Chief Meteorologist, U.S. Department of Homeland Security, who shared that one of the largest challenges in her work has been communicating extreme impacts within the current classification system. For example, a tropical storm with atypical hazards is dismissed as being "just a tropical storm, so I'm not going to stand up an EOC." She also noted that traditional metrics may miss the atypical features of an extreme storm. Wescott asked whether agencies are considering changing the categories themselves, and how storms are classified, in addition to or instead of changing perception of the categories. Senkbeil responded that the National Hurricane Center (NHC) is ultimately in charge of classification decisions and determining whether such changes would yield more accurate perceptions of risk and improve responses. Senkbeil added that the one-to-five scale category is a useful but very simple product and that Schauer's graphics are another successful, but more complex, way to quickly communicate the different hazards and uncertainty. Calling for products that find the middle ground, Senkbeil noted that most people spend about 20 seconds getting information. Moulton responded, saying that Wescott helpfully foregrounded the idea that people tend to miss impacts from "background level" hazards that do not fit or meet the criteria of in-place classification systems. She mentioned the Waffle House Index—based on the restaurant's reputation that it remains open in all but the very worst storms— as an informal measure that is quite effective at helping people to understand the challenges they are facing.

Regarding the concept of cascading or compounding effects, Castle Williamsberg, Social Science Research-to-Applications (R2X) Coordinator, NOAA, noted that "it's crucial to also evaluate the effects of our risk communication messaging across our information ecosystem" and raised the question of how best to evaluate these communication consequences. For example, Lindner noted the cascading impacts of a loss of power, which can result in loss of communications channels, emergency services, and back-up power. The question of whether forecasters discuss impacts, particularly cascading impacts, is particularly important, he said, because people's perceptions about the accuracy of the forecast are sometimes based on their prior experience of impacts.

HIGH-LEVEL SUMMARY OF SESSIONS ONE AND TWO

Andrea Schumacher, Ann Bostrom, and Hugh Walpole, Associate Program Officer, National Academies of Sciences, Engineering, and Medicine, provided a high-level summary of the first two sessions. Many of the takeaways wove throughout the workshop, including the central importance of strong partnerships, the usefulness of localized information alongside general forecasts, the challenges inherent in communicating about complexity and uncertainty, and the importance of listening to target audiences.

A major challenge to risk communication is the noise that is produced by a glut of information that does not help individuals understand what they might expect and how they might prepare (i.e., historical facts, general data about category or intensity of a tropical cyclone). Schumacher explained that in this context, noise can be described as statements and graphical information "that don't necessarily relate to locally experienced impacts." This challenge dovetailed with another common theme, that is, whether and how "impersonal graphics [i.e., not tailored to any specific community] convey personal risk."

Bostrom noted a similar theme in "all events are local," in terms of both localized information sent as part of the messaging and local data about actions taken by individuals in response to that information and their own experiences, gathered through social observational research. Understanding both the specific local impacts of a given hazard (which is often one of several hazards) and the bigger picture of how a multi-hazard storm evolves and moves is important. Such an understanding involves tracking the dynamic nature of the storm, as well as the speed and location of impacts of multiple hazards happening at the same time. This need for agility in storm tracking has a counterpart in the need for agile, on-the-ground research before and during the event, supported by a strong research infrastructure. Tracking what people experience, and how they act in response, can deepen understanding of how they will respond in the future, which ties into the concept of benchmark storms.

As storms are becoming more complex, communication tools are becoming more sophisticated, and this trend. Bostrom linked this theme to two important topics from the sessions. First, more research is needed on how information moves across various platforms (within both the public and private sectors) and on how to leverage artificial intelligence (AI) tools to communicate more effectively. Second, and relatedly, in the face of evolving climate change and in "the context of a new normal of increasing extremes" the need to be nimble and consistent is even more pressing. This need can be met through strong partnerships with good information flow between partners, especially in the face of extreme or unprecedented storms. A "strong coordinated community" is critical to consistency in messaging that goes beyond "traditional" messaging.

Finally, Walpole noted a persistent issue that centers on categorizing a compounding or cascading disaster. Categorization decisions include how to communicate about complexity and uncertainty and which hazard to focus on in a multi-hazard context. Therefore, Walpole noted, panelists grappled with the ethical dimensions of these decisions, including who decides the details and who creates the messaging.

SUMMARY OF BREAKOUT DISCUSSIONS: APPLYING RISK COMMUNICATION LESSONS FROM OTHER HAZARDS TO THE TROPICAL CYCLONE CONTEXT

Breakout discussions about applying risk communication lessons from other hazards—earthquakes, extreme heat, flooding—to the tropical cyclone context followed the keynote and panel discussion. Jeanette Sutton (committee member), Associate Professor, College of Emergency Preparedness, Homeland Security and

Cybersecurity, University at Albany, State University of New York (SUNY), introduced this portion of the workshop. She explained that the main objectives for this session were to mine past risk communication experiences for insights into future endeavors, and to discuss how the scope of research around tropical cyclone risk communication might be expanded.

Earthquakes

Richard Allen, Director, Seismology Lab, University of California, Berkeley, shared the takeaways from the breakout discussion on earthquakes. He began by stating that earthquake early warning efforts in the United States are relatively recent and successful. The U.S. Geological Survey (USGS) is the agency responsible for issuing the alerts that are part of the early warning system, from monitoring and reporting on earthquake activity to warning people. With this system now in place, the opportunity exists to evaluate the effectiveness of message delivery—how to make messages more impactful, how to reach more people, and how to support more effective action. The evaluation should consider the roles of different channels (e.g., Wireless Emergency Alerts [WEA], Android apps) and of the speed of onset. A 10-second warning, for example, does not allow for either elaborate messaging or action. There is "an interesting tradeoff between the . . . lack of warning time and the simplicity of the message." A simple message may help more people take more effective action, Allen explained. He commented that including maps in messages is not effective because understanding them takes more time than people usually have. Very short warning times necessitate education and preparedness before the event to help people know how best to understand messages and take immediate action. A third aspect of effective communication is the use of clear, understandable language and concepts; Allen highlighted the term "intensity," which is of critical importance to the warning community, such as forecasters, but largely misunderstood by (or unknown to) the public.

Extreme Heat

Micki Olson, Senior Risk Communication Researcher and Project Manager, Emergency and Risk Communication Message Testing Lab, University at Albany, shared the takeaways from the breakout discussion on extreme heart. She commented that extreme heat has no standard definition, which can lead to inconsistent messaging across local and state-wide agencies (e.g., how watches and warnings are issued). At the same time, extreme heat affects millions of people and has a "public health component that we don't usually see with other hazards." Research has shown that equity and the vulnerability of different populations make how risk is communicated especially important: demographics and, to a lesser extent, location have been shown to affect individuals' risk perceptions and the extent to which they are able to heed heat warnings and take appropriate action. Jargon, technical language, one-way communication from expert to public, and assumptions about what leads to behavior change all inhibit people's understanding of how extreme heat might affect them. However, most people do understand that they are vulnerable and "generally understand the benefits of what they're supposed to do to protect themselves."

Further, experts do not usually understand how their messaging is received by the public. People might not understand the specific differences between, for example, a tropical storm, tropical depression, hurricane, or tropical cyclone. Olson highlighted the group's focus on plain language, localized messaging, and consistent messaging as critical for effective communication, along with increased attention to the public health dimension of extreme heat.

Flooding

Amanda Schroeder, Senior Service Hydrologist, National Weather Service, reported on the third group's discussion about flooding. Coordination between the private and public sectors, as well as the need for stronger cross-sector relationships, especially between academia and government agencies, was a focus of that discussion. One example of coordination is flood inundation mapping by the NWS and nongovernmental entities, such as Flood Vision, a private-sector tool by Climate Central.[4] Schroeder noted that different sectors often want different information, with different needs (including the role that timing plays in communications practices) and different content (e.g., messages that contain model-based, deterministic information versus probabilistic information). Discussants also raised the idea that the government is "often not at the forefront of technology advances," a standing that might be improved through better partnerships.

---

[4] More information about Flood Vision is available at https://www.climatecentral.org/floodvision.

Chapter 4
Risk Communication and Decision Making in Communities

Messages about various hazards can often come from multiple sources during a single event, and the many decision-makers involved are often responding to or addressing needs and pressures specific to their situation and audience. Craig Fugate (committee member), Craig Fugate Consulting LLC and former Administrator of FEMA (2009-2017), articulated this idea in his introduction to the first panel of session four. Session four focused on risk communicators, community leaders, and decision-making around risk communication, especially within the community context. The goal of the session was to understand various risk communication needs and sources across scales and communities (e.g., county, municipal, faith-based organizations, local emergency management) and challenges that arise across population segments with differing experiences. The session included a panel and a roundtable. Session speakers illustrated how decisions about how and what to say, and when, can profoundly affect how the audience receives the message and how they respond.

RISK COMMUNICATION ACROSS SCALES: RISK COMMUNICATORS IN COMMUNITIES

Amid all the effects and impacts of extreme weather events, "the only thing that matters at the end of the day is how many people lost their lives that we could have prevented," said Fugate, moderator for the panel. He shared that this observation was rooted in his interest in Hurricane Ian, a storm in which "more people were drowned or killed from blunt trauma in evacuation zones than any hurricane in Florida" since 1935. The central importance of lives impacted echoed throughout the following presentations is in service of understanding "the various risk communication needs and sources across [varied] scales of communities," as Fugate stated, including at the household, municipal, county, and state levels. Drew Pearson, Emergency Management Director, Dare County, North Carolina, the panel's first speaker, stressed the personal nature of this work: people who have lost their lives in these events were known personally at the local level.

Pearson discussed the role of risk communicators working at the local level, "where it starts and ends." Local-level decisions around risk communication are highly contingent on multiple factors, including historic data on the impacts of water (e.g., feet above ground level) and wind (e.g., sustained strength over an extended period of time). These data help the local emergency management office predict when present conditions might become unsafe and allows them to begin messaging efforts early. Forecasts from the National Weather Service (NWS) also factor into decisions about when and what kinds of risk communication products are issued, and what kinds of actions are recommended or mandated. The warning coordination meteorologist at the NWS is an important partner for local decision-makers. During weather events, there is often near-constant engagement between local government offices and the regional NWS office in Newport/Morehead City, North Carolina.

Another important stakeholder in making decisions and then communicating subsequent information and instructions to the public is the county public information officer within the public relations department. Pearson emphasized the importance of preparing beforehand, through community preparedness forums that seek to educate the public about specific terminology, a storm surge awareness campaign, and building strong relationships between the local government office and the media, businesses, local influencers, and the public in general. Furthermore, the public information officer is not "just pushing information, they are listening" and working to bring community issues to the county staff before those issues grow into "hot button topics with our elected officials." The county public information officers, through their work to build this close working relationship, "not only ensure accurate information is being shared, but also allows messaging to be adjusted to answer questions or concerns that [these officers] are hearing, and to correct misinformation, whether it's shared deliberately or unintentionally."

Pearson also reviewed some in-the-moment strategies for risk communication adopted by Dare County. These strategies include graphics, public safety announcements, door-to-door communication, and short videos pushed through a mass notification tool called OBXAlerts,[1] as well as making use of the Federal Emergency Management Agency (FEMA)'s Integrated Public Alert Warning System (IPAWS).[2] With IPAWS, Pearson noted, the messaging changes from informing the public of the possibility of hazards to "clear, concise details about the hazard," including what exactly it is, where and when it will occur, and how to stay safe.

Pearson closed his presentation by emphasizing qualities that often define the emergency manager (EM) role: a deep knowledge of the local community, cross-sector partnerships, connections made across the timeline of an event (e.g., before, during, and after), and persistence in both communicating risk and seeking ways to improve risk communication. He noted that getting people to understand their

---

[1] More information about OBXAlerts, a platform for alerts and notifications from Dare County and surrounding towns, is available at https://www.darenc.gov/departments/emergency-management/emergency-alerts.

[2] More information about IPAWS, a national system for local alerts, is available at https://www.fema.gov/emergency-managers/practitioners/integrated-public-alert-warning-system.

risk is always a challenge, but that continuing to improve communications skills is important and always ongoing.

Russell Strickland, Secretary for Emergency Management in Maryland, then discussed decision-making at the state level. People are the focus of all efforts, which are aimed at improving chances of survival, he noted, echoing Pearson. The state of Maryland works within a tiered system, in which planning, response, and recovery efforts unfold at the local, regional, state, and federal levels. "All events are local," and thus, in this "across-the-board team effort," the state's primary role is to support local jurisdictions as they respond to threats and hazards in their communities.

Strickland highlighted three key aspects of the state's role in decision-making around risk communications and emergency response. First, issuing information about and warnings for all hazards to the public is "a core capability in all phases of emergency management" (e.g., mitigation, preparedness, response, recovery). In Maryland, Strickland notes, this capability exists at both the local and state levels of government, the latter of which supports the former in communicating to hazards and threats. Second, the state coordinates communications and responses when individual jurisdictions are overwhelmed or when multiple jurisdictions are involved and affected. Finally, the state is responsible for delivering "coordinated, prompt, reliable, and actionable information [about a hazard or threat] to the whole community through the use of a clear, consistent, accessible, and culturally and linguistically appropriate method." The goal is always to support local jurisdictions in their responses by "provid[ing] the right information at the right time to the right people so they can make the best decisions," Strickland emphasized.

EM decision-making is shaped not only at various levels of government, but also by specific threats and hazards. In the latter context, Daphne LaDue, Senior Research Scientist, University of Oklahoma, and Tom Cova, Professor of Geography, University of Utah, discussed tornadoes and fire, respectively. LaDue spoke about her research into risk communication around tornados between city and county EMs, other local officials (e.g., fire captains, school officials, public works supervisors), and survivors of tornadoes. She stressed that most communities and individuals within a tornado warning area will not directly experience the tornado. Thus, individuals will often not take protective action until they actually see, hear, or feel a tornado coming; and EMs on the other hand, by the nature of their responsibilities, take action but also look to minimize the cost of being prepared. "The hard truth [for EMs] is that their jurisdiction probably won't get hit." Finances play into response decision-making, she said. Launching a tornado response is costly, and LaDue pointed to one community that had spent $200,000 in overtime and other forms of preparation but did not have a single tornado touch down. Past experiences also play a role in decision-making, especially as EMs seek to narrow down when and where impacts might happen within the 2- to 8-hour forecast window provided by the NWS. The timing of forecasts and warnings from the NWS also influence how much time EMs have to make decisions in order to get personnel in place if needed.

Research by the University of Oklahoma's Institute for Public Policy Research and Analysis builds on LaDue's studies, she noted, and includes a survey asking EMs to rank what type of information was most important at key points in a severe

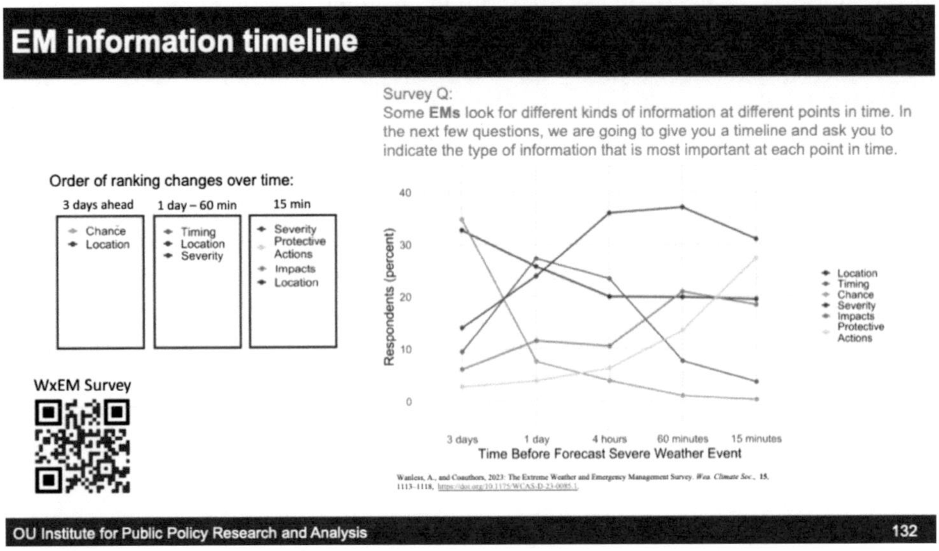

EM information timeline

Survey Q:
Some **EMs** look for different kinds of information at different points in time. In the next few questions, we are going to give you a timeline and ask you to indicate the type of information that is most important at each point in time.

Order of ranking changes over time:

3 days ahead	1 day – 60 min	15 min
Chance	Timing	Severity
Location	Location	Protective Actions
	Severity	Impacts
		Location

WxEM Survey

OU Institute for Public Policy Research and Analysis — 132

FIGURE 4.1 Survey results from EMs ranking what type of information was most important at key points in a severe weather event: 3 days, 1 day to 1 hour, and 15 minutes before the forecasted storm.
SOURCE: Wanless, A., S. Stormer, J. T. Ripberger, M. J. Krocak, A. Fox, D. Hogg, H. Jenkins-Smith, C. Silva, S. E. Robinson, and W. S. Eller. 2023. Th e Extreme Weather and Emergency Management Survey. Weather, Climate, and Society 15(4):1113-1118. https://doi.org/10.1175/WCAS-D-23-0085.1. American Meteorological Society. Used with permission.

weather event—3 days, 1 day to 1 hour, and 15 minutes before the forecasted storm. The importance of information about location, timing, chance, severity, impacts, and protective actions changed over time: 3 days out, chance and location were the top ranked types; between 1 day and 1 hour, timing, location, and severity were the top three types; and, 15 minutes out, severity, protective actions, impacts, and location were listed as the most important types of information needed (see Figure 4.1).

LaDue reported that her own research shows that in the southeastern United States, where tornadoes are notoriously difficult to forecast, EMs have developed an "amiable distrust" of forecasts. Although they trust the forecaster's intent, she found, EMs do not trust the forecast itself, because of the number of times a forecast involves either a null event, a short- or no-notice event, gaps in information, and/or differences in perspectives. As a result, LaDue found, EMs often rely on their own past experiences and judgment to "adapt to the shortcomings in the state of the science" and to calibrate their responses to the forecast based on their own experience-informed perception of risk. This strategy might mean that they remain even after the Emergency Operations Center (EOC) has closed or they wait to activate emergency response teams. Such calibration is anchored in learning, which is itself the combination of experience and reflection on that experience, LaDue explained. This learning process could be encouraged through conversations like a post-event review between local officials and constituents that allowed officials to explain decisions and contribute to a deepening of trust (see Olson et al., 2023).

Many approaches within the decision-making system around tornado risk are working, LaDue added. EMs serve as an important conduit for information, gathering critical and accurate data and passing it along to other local officials. The relationship between EMs and the NWS is strong, which is critical to the flow of accurate information and weather-related decision-making.

LaDue concluded her presentation with a brief discussion of gaps and opportunities in this area. She noted that demographic information of EMs as a group is lacking, as are data about job stability and turnover. In addition, the current forecast scale—which covers a relatively large geographical area—does not serve the small geographical area of concern for an individual or a city or county EM. It would be helpful to learn what forecast information means for them and use this knowledge as a basis for some level of autonomy as they make decisions for their communities. Figuring out how to support them in this learning and decision-making is important. Finally, there is a need to build greater resilience to severe weather, largely because many decisions are made in the final moments before a tornado strikes, and actions are not taken until the threat is clear.

Tom Cova focused his discussion on wildfire risk communication and the timing of decision-making and evacuation processes. Elements of timing include lead time (i.e., the duration between the trigger point and when the situation becomes dire); the length of time it takes for officials and then the public to make decisions; and the design of the community itself, specifically the paths of egress. Delays or mismatches in these elements of timing can make a situation dire; for example, in the Camp Fire of 2018, the time to evacuate exceeded the lead time, so the situation was dire from the start.

Exurban development at the periphery of metropolitan areas can cause longer evacuation times, Cova said, which can make a situation dire before people even realize it (Cova et al., 2021). Fires have also become more extreme because of climate change, which brings extreme winds and greater drought that can affect spread rate and flame lengths in unprecedented ways. Uncertainty exacerbates the problem of timing because it makes it extremely difficult to accurately predict the time available to evacuate. Cova emphasized how dynamic these situations are: the fire can spread more quickly than anticipated, or in a different way than anticipated and roads can close or become blocked. Cova also pointed to situations when little to no warning was given because warning systems were not activated. He refers to this situation as a "human problem" wherein a clear trigger point was not set, or it was not set at the right point. In these cases, challenges around the timing and sending of alerts has led to longer evacuation times in dire situations.

Echoing other panelists, Cova emphasized the uncertainty that arises particularly around unprecedented events. Models are often too optimistic, he said, and do not simulate dire enough situations. All of these factors—uncertainty, more extreme events, exurban development, and the need to simulate more dire events—highlight the "need to assess and improve protective-action risk communication."

Social inequities also play a role in the effectiveness of risk communications. Jim Elliott, Professor of Sociology and Co-Director of the Center for Coastal Futures & Adaptive Resilience, Rice University, spoke about how social factors influence how people understand and respond to risk more than researchers might think. For example, a failure to evacuate is often explained as a transportation gap—that is, when people do not have access to a car or cannot drive because they are too young or otherwise unable to drive. This transportation gap can be directly addressed by increasing access to transportation in areas with high concentrations of people without cars and by creating shelter-in-place plans that, perhaps, use public infrastructure for worst-case scenarios when people cannot evacuate. However, lack of a vehicle or access to transportation is just one reason why people may not heed evacuation orders, he explained.

Elliott's research has revealed that a person's decision to remain in place may be explained by three sociological reasons: religion, race, and roles (see Haney et al., 2010). For individuals whose religious beliefs explain extreme weather events as acts of God, remaining in place can be an act of trust in God. In this case, effective risk communication may entail working more closely with religious leaders, who may be more trusted sources of information than government sources. Such leaders could encourage their congregations to interpret and respond to such events in ways that align with and amplify government messages and mandates.

Racial inequities can increase exposure and vulnerability to hazards and threats in a multitude of ways, Elliott said. One less-discussed area of vulnerability is that homeowners of color often have lower wealth and less comprehensive insurance. In this case, an inequitable recovery process makes a quick and complete rebuild difficult, which in turn informs planning before and during a storm. Owning a home creates a strong incentive to stay and repair damage as quickly as possible. Concerns that a small problem will become a big problem if not fixed immediately can lead to a person remaining in place rather than facing of potential for leaving their home and being unable to return. "People sometimes stay precisely because the risk that's communicated creates a lot of worry about what's going to happen to their home and their economic future," Elliott noted. Situations such as these are particularly common in historically marginalized communities, where not only are resources scarce but also trust that government assistance during the recovery will be adequate is low. Elliott suggested that government communicators could help to address this issue by engaging more with economically vulnerable communities, particularly in areas with high homeowner rates, where people are incentivized to stay and protect their financial investment.

Household roles also influence how people respond to an evacuation order. Elliott noted that often, traditional gender roles come into play, with women and children evacuating and men staying behind to protect the property. In addition, household earners who are worried about losing their jobs often stay behind. In these cases, engaging with employers, particularly those who employ "hourly workers in traditionally male sectors," could help to improve response to evacuation orders and similar types of messages.

Messaging about the hazards themselves is not enough, Elliott noted. The cascading and compounding effects that impact the social and built environment also must be communicated to the public with an eye toward social factors. He cited the example of road closures due to street flooding, which can cut people off from important services even though they are not in a flood zone. Another example is the threat posed by hazardous industrial pollutants that could be released if infrastructure is damaged during a storm. Elliott suggested that people responsible for risk communication might prioritize areas threatened by both chemical and natural hazards—for example, by engaging in early and ongoing communication with local leaders to raise community awareness about how the mobilization of hazardous chemicals via local storm waters can make remaining in place more dangerous than it might seem based on the storm forecast alone.

Fugate, as moderator, opened the question-and-answer portion of the session by observing that there is a legal distinction between warnings, which are issued by the NWS, and evacuation orders, which can only be issued by state and local jurisdictions. Therefore, he wondered, who makes these decisions, and what are their sources of information around those decisions? Pearson explained that in North Carolina, the authority to issue an evacuation order, declare a state of emergency, and impose various prohibitions or restrictions lies with the governor, who delegates to county officials, who then may delegate to municipal officials. Thus, one challenge for local EMs is drawing together the NWS forecast with local knowledge and ensuring that elected officials at the county and local levels have correct, critical information as they decide on the course of action. State-level decision-making resembles the local level in terms of briefing officials in order to help them make informed decisions, Strickland noted. Discerning what information officials are looking for as they make decisions is critically important, he noted. In Maryland, he explained, in the event of a storm, the state names an evacuation coordinator who works closely with officials (e.g., county manager, mayor) in the local jurisdictions that are most likely to be impacted, and across jurisdictions as well, and keeps the governor apprised.

Rumors and misinformation—whether deliberate or unintentional—pose significant challenges to risk communication and response, and were the topic of Fugate's second question: How can rumors be addressed and controlled? Listening to what sort of messaging and information is circulating is key to fighting misinformation and rumor, Pearson said. His county establishes a call center with a hotline that people can call to receive accurate information and report any rumors that are circulating. With this knowledge, county officials can directly counter misinformation with accurate facts from a trusted source. However, Pearson stressed, officials must be tuned in and listening to what the public is communicating. Strickland added that similarly, in Maryland, smaller jurisdictions establish rumor control hotlines, which feed into the Joint Information Center, which then verifies information to send back to the jurisdictions.[3] Elliott noted that a large part of rumor control for his jurisdiction is coordinating consistent messaging among the many agencies commonly involved. In addition, alerting local news outlets to prioritize or use information coming from a central source, rather than other sources of potential misinformation, ensures that they are "a partner in these conversations." Because many people turn to the local news when storms and other hazards threaten, coordinating with local news outlets can be especially helpful, he said.

Castle Williamsberg asked the final question, "What is the single-most important operational challenge faced by practitioners in effectively communicating

---

[3] Information about a Joint Information Center from Washington County, Maryland, is available at https://www.washco-md.net/jic/.

weather-related risks? It is a challenge to keep the message simple, Pearson said, but it is essential to be "clear, on point, and unambiguous." Strickland agreed, adding that balancing uncertainty and confidence in the midst of a changing forecast is also challenging. Giving people accurate, meaningful information with enough time for them to respond can be difficult. "Understanding what people are responding to" can be essential, LaDue added, which might include awareness that multiple warnings or hazards (e.g., hail and winds) are driving behavior, depending on which is prioritized. She stressed the importance of understanding how the communications and decision-making systems work: a warning from a local official (e.g., a sheriff) may feel more directed and relevant than one from the state to a larger or more general area.

ROUNDTABLE: COMMUNITY LEADERS AND COMMUNITY ACTION

The session on risk communication and decision-making in communities continued and concluded with a discussion between Jeff Lindner and Archie Chaisson, Parish President, Lafourche Parish Government, Louisiana, chaired by Brad Colman (committee member), President, American Meteorological Society. The two discussants began by talking about the information they consider when making decisions about risk communication in the event of a storm, and what information they would like. Lindner stressed the importance of "primary sources"—the NWS and the National Hurricane Center (NHC)—in gathering information with which to brief officials during a hurricane or other event. The information from these two sources was important, Lindner said, in guiding decision-making during Hurricane Laura (2020). While TV meteorologists and European weather models made more disastrous and different predictions, the NHC's forecast "really never changed for about 48 hours," and the NWS expressed a high level of confidence in that forecast track. In this case, basing decisions solely on information from the NWS and NHC resulted in an appropriate level of response and not an overreaction. "It's just as important sometimes not to pull the trigger as it is to pull the trigger" Lindner noted, especially when dealing with a population size in the millions, which can be the case for storm surge impacts and warnings.

Lindner highlighted the mismatch between various milestones on the timeline for hurricanes at which decisions must be made (to allow enough lead time) and the timing of when crucial information is available. Ideally, he said, the lead time on the forecasts from the NWS would be longer. Because of the current mismatch, decisions must be made even when the level of uncertainty in high: for example, in Harris County, Texas, mobilizing busses for use in an evacuation takes about 96 hours, and thus the decision to order this resource from the state must be made 96 hours out, when uncertainty is often high, if it is to be effective.

At 60 hours out, Lindner explained, officials on a multi-agency call decided to evacuate the general public, a process that involved coordination among counties or parishes, including, for example, a phased approach so that coastal counties issue evacuation orders several hours before more populous metropolitan areas such as

Harris County, so that the coastal population can clear out before the traffic increases. This approach is especially important, he noted, in high-population areas such as the Texas coast.

In addition to population density, lag time between order and response is another important factor to consider when contemplating the timing of messaging, Lindner added. There can be a 12- to 24-hour gap once decisions about evacuation become public. Closing the schools once the evacuation order has been given can help shorten this lag time, he said.

Uncertainty can profoundly impact the decision-making process, including questions of how to communicate uncertainty itself to stakeholders, whether that be the public, elected officials, or the media. Chaisson commented that uncertainty often looks like preparing for something that does not actually happen—i.e., "crying wolf." This situation can happen repeatedly, as in the storm season of 2020, and thus skew people's sense of risk in the event of a truly threatening storm. Similarly, people will remember a benchmark storm that they survived and disregard the present threat. These two groups of people are more likely to remain despite evacuation orders, and it can be very difficult to effectively communicate uncertainty to them. Chaisson referred to Hurricane Ida (2021), which was an unprecedented storm for the area, and described how he and others worked to communicate that this event would in fact directly hit the area. He noted that because Hurricane Ida was more of a wind event than a water event, the impacts were not as catastrophic as they could have been. Lindner briefly added that a confident forecast from the NWS or NHC helps to remove uncertainty, which is reflected in more strident language used by both the NWS and the local EMs.

Costs associated with evacuation orders often impact decisions about whether or not to issue them, especially when uncertainty is high, Colman noted. Cost is secondary to saving lives, Lindner stated. However, the dampening effects of alert fatigue are problematic as well, and striking a balance between the importance of saving lives and the necessity of avoiding alert fatigue can be difficult. Normalizing situations where "nothing happens" can be helpful, he suggested. Chaisson noted the way that cost factors in the decisions of individuals who cannot afford to evacuate: "they don't have the means, and they won't leave for nothing." Certainty has to be high for this population to evacuate.

Colman stated that "environmental equity," a term used in relation to populations who are more vulnerable because of their location and/or lower income, is also an important factor in decision-making in the moment, as well as for the future. Chaisson noted that, in their parish, evacuation protocols and other plans are based on proximity to the coast; the population deep in the parish, closest to the coast, is environmentally more exposed to threats and experiences a higher rate of poverty. Evacuation often involves moving people up the coast and parish little by little, rather than all at one: people will first go up to family and friends, and then slowly "migrate up," often staying in hotels or schools serving as temporary shelters. This approach differs from mass evacuation, which was necessary for Hurricanes Ida, Ian, and Harvey, which brought flooding to densely populated areas and necessitated chartering busses and erecting mega-shelters.

Educating the public is key, Lindner added. Education can involve working closely with local leaders, especially church leaders, before storms to help people understand how to navigate the evacuation system. It requires clarity in the messaging about the type of evacuation response necessary given the various threats and hazards. Lindner cited the example of people believing they need to drive very long distances to escape a hurricane when, in fact, driving a much shorter distance to escape storm surge flooding may be all that is necessary. Distributing information ahead of time is good, but the underlying challenge is that "no one really pays attention to this until it's happening." Chaisson and Lindner both reiterated the challenges to communicating risk to people who have experienced evacuation orders when "nothing hit" and to people who stayed and survived bad storms in the past—and, in both cases, downplay the risks of the present threat.

Chaisson noted that decisions about evacuation are never simple or easy. He particularly stressed the emotional difficulty of ordering an evacuation—essentially asking a person to leave everything they have with no assurance that they will be able to return to it, or that it will be there if they do. It is "probably one of the most difficult things you will ever have to do." He emphasized that often leaders need to trust their gut, prior experience, and the plans already in place.

HIGH-LEVEL SUMMARY OF SESSIONS THREE AND FOUR

Jeanette Sutton, Craig Fugate, and Brad Colman provided a high-level summary of sessions three and four. Themes and concepts that cut across the two sessions—and resonated with the broader workshop discussion—included the need to use plain language and avoid jargon; the crucial role of partnerships in deepening understanding of the various stakeholders' needs; and the importance of considering those needs when building and implementing a communications strategy, crafting messages, and making decisions. They acknowledged that any given audience or population is not monolithic; therefore, communication strategies would ideally reflect audience diversity.

Sutton noted that discussants reinforced the concept that messaging works best when it also addresses a specific audience. Here, messengers must understand who the members are, what they need, and how they understand different types of information. Whether that means a map, a focus on impacts, or an emphasis on vulnerabilities, attention to the audience's greatest needs complements the use of plain language. One example of an audience requiring tailored messaging is people who are undocumented, given current political realities. The two-sided nature of communication—message and receiver—was raised in discussions about new technologies and products, which could be created and assessed with audience needs in mind.

Although the hazards discussed by the breakout groups—earthquakes, extreme heat, and flooding—vary greatly in terms of speed of onset, lead time, interaction of compounding hazards, and partner organizations that can serve as intermediaries between government agencies and the public, they offer common themes in terms of risk communication. One shared theme is the importance of using plain language, rather than jargon or overly technical language. Fugate shared the example of the term "storm surge," which is often not accurately understood by the public. The "tendency for precision" can be a hindrance in helping individuals understand how to act.

Fugate, reporting on the first set of panels from session four, echoed Sutton's highlighting of attending to the different needs of the community members that comprise an audience. In this, he captured a point made throughout the day: that messaging often works best when it addresses a specific audience. This involves understanding who that audience is, what they need, and how they understand different types of information. He highlighted one phrase— "We trust the forecasters, not the forecast"—as being particularly revealing of how the general public responds to uncertainty. He raised the question of how to communicate particularly with people who are undocumented, given current political realities.

Another audience, as it were, is EMs; Fugate highlighted the discussion around the need to better understand how and why EMs decide to activate or not activate warning systems, especially when the situation involves high levels of uncertainty. Coleman noted that EMs approach the decisions and pressures they face with a great deal of preparation, ingenuity, and resourcefulness, helping them to be focused and decisive in the moment. Discussions also highlighted the great effort and wide range of resources that can be used to mobilize an emergency response. In all of this, Colman noted, strong partnerships—between local EMs and the NWS—help EMs to filter out the noise. Finally, EMs sometimes advise, or are themselves, elected officials; therefore, their decisions could become an election issue—which resonated with the workshop-wide focus on the local dimensions of information and communication.

Partnership emerged as another prevalent theme and spans connections across the public and private sectors (especially around technology), as well as intermediaries such as community organizations that can help connect researchers and government agencies with locals and public health organizations.

Chapter 5
A Recap of the First Day of the Workshop

Rebecca Morss, Program Director, Office of Integrative Activities, National Science Foundation, provided an overview of day 1 to set the stage for day 2. She first synthesized key themes from the panel discussions and then highlighted some common overarching themes. Morss began with a point made frequently during day 1: every event is different. The details of the event itself, the location, the community, and the meteorology all serve as different variables that determine impacts. Two major elements identified by Morss within this list are the built infrastructure and people's responses to warnings and other information about the storm. Although the data on the impacts of various hazards on the built environment are robust, the data on how people respond—especially data collected in near-real-time—are much less plentiful. Gaps include evaluations of what information (including risk warnings) people receive, how they understand it, and what actions they take in response. Real-time research in this area received special emphasis, Morss noted.

Morss connected this need for data on people's responses with another common theme, the many benefits of multiple perspectives. Often the problem of the storm, as it were, is so big and complex that the whole cannot be adequately viewed from a single perspective; instead, people perceive the storm in many different ways, depending on their roles, needs, and location, and multiple perspectives are useful for crafting effective messaging that enables responses. Multiple perspectives are especially critical in the fundamental task of defining the problem or goal—another common theme highlighted by Morss. "How we define a problem influences how we solve it," she noted. Clarifying the problem is particularly important when multiple perspectives come together. For some speakers, including emergency managers (Ems) and other officials, the ultimate problem was how best to reduce death in the midst of an event; this perspective was largely influenced by their role and responsibilities. For others, in other roles, the problem might lie in clearly defining and characterizing the various kinds of storms because extreme events have become the new normal. Morss wondered whether using the label "extreme" really matters; rather, simply determining whether people are at risk may be most important. The work of characterizing the problem is especially urgent in the case of extreme storms, when defining and managing risks can be difficult given the lack of precedent, she noted.

The question of how best to communicate about the risks involved in multi-hazard storms, as well as complex and cascading hazards, was a common theme from day 1. Morss referred to discussions about how hazards impact people directly and indirectly (e.g., disrupt access to food or other basic needs); "...the hazard is just one part of a bigger system," she noted, and its effects can be wide-ranging.

The theme of uncertainty was present in discussions on a variety of topics. Meteorological uncertainty and spatial variability—and how to communicate probability to the public (e.g., cones of uncertainty)—is one example. How uncertainty intersects with decision-making was also a frequent topic. Morss particularly highlighted how attending to multiple perspectives seemed to reveal uncertainty as relative.

The terms "vulnerable," "storm surge," and "cone of uncertainty," as well as the nature of "traditional" messaging served as touchpoints in discussions about using plain language and avoiding jargon. Morss summarize the main questions: Do people understand what the terms mean? Do they understand what is recommended, and why that can help?

Morss illuminated time as an important theme that was discussed in several different ways over the course of the day. Timing is a critical factor in decision-making and action, including giving people enough lead time and considering factors that cause people to lose time in which to make decisions and take action. Related issues include the time people need to process information, to absorb the information's pertinence to them and the need to take action, and to then decide on the action(s) (including waiting, which is itself an action). The shift from "this is happening" to "this is happening to me, and I need to do something about it" takes time. Morss added that the temporal dimension of events challenges forecasters' decision-making processes. She recalled descriptions of situations when officials and members of the public were surprised by the direness of the situation, which in turn eroded the time in which people had to make decisions and take protective action.

The theme of trust—who to trust, how to build trust—was also part of the first day's discussions, Morss said. Trust among partners, and trust between agencies and members of the public, are important for effective risk communication. Efforts to build trust should occur not only between events—in the shape of longer-term relationships between partners and communities—but also during events. Also important is the trust people have in the forecaster—over and above the forecast itself. She wondered about the interaction between this trust in human intelligence and presence, on the one hand, and artificial intelligence (AI), on the other, and posed the open-ended question of "Does trust in the human element shape trust in AI?"

The final theme from day 1 that Morss highlighted was the strengths-based view of the past few decades: many advances have made things better for so many people. She referred to "storm surge" as an example of a confusing concept that has become more accessible; through the combined efforts of many stakeholders—including the National Weather Service, the National Hurricane Center, broadcast meteorologists, and EMs—improvements have been made in how the risks of storm surge are both predicted and communicated.

Of course, opportunities remain, Morss said; even as one problem is solved, others appear. Understanding of the interrelationships between different hazards might be leveraged to help people take appropriate action. For example, extreme heat might follow in the aftermath of a hurricane because of power outages and lessons from heat waves could be applied; or last-minute decision-making during an earthquake might be applied during a flash flood. In addition, messaging can become complicated in extreme events that surpass any benchmark storm. Comparison, in this case, might be leveraged to highlight what we do not know, Morss noted.

Another cluster of open questions involves evacuation choices. Why do people do what they do, when they do? Morss linked this behavior back to timing, including the need to help people who chose to evacuate to get to their destination in time to stay safe. She emphasized the importance for real-time research, which might yield a deeper understanding of why people leave when they do—what was it about that moment that spurred them to leave? Because people are influenced by different things at different times, it is important to communicate "across the lifetime of a hazard."

Chapter 6
Practical Translation of Risk in the Public Arena

"How do we determine how to best bridge research and practice to advance risk communication?" was the first question posed by Gina Eosco, Division Chief of the Science, Technology, and Society Division, National Oceanic and Atmospheric Administration (NOAA), during her presentation on public initiatives to evaluate the effectiveness of risk communication products. This question encapsulated the theme of the second day's first session: Practical Translation of Risk in the Public Arena. Panels on the work of translating risk from research to practical messaging and responses included discussions of risk communication innovations and new frontiers in communication around tropical cyclones in the public and private sectors, as well as descriptions of specific communications technologies developed by private companies.

RISK COMMUNICATION INNOVATIONS AND NEW FRONTIERS IN TROPICAL CYCLONE COMMUNICATION: PUBLIC SECTOR

The efficacy of technologies and approaches that facilitate communication of risk—and uncertainty—to the public would likely benefit from being evaluated. This panel, composed of representatives from the National Hurricane Center (NHC) and NOAA, addressed this topic from the public-sector perspective.

Andrea Brennan, Director, NHC, began the panel by discussing the work of communicating risk at the national level in the event of a tropical cyclone. The NHC tracks the entire "lifecycle" of the event, which means that risk communication often spans multiple hazards that vary in severity across time. This work requires the coordination of simultaneous different messages to multiple populations in multiple locations, because experiences of stages and impacts differ at any given point in time and by location.

The timing of risk communication is an important factor in helping individuals and communities take protective action. Risk communication faces extra challenges around "low-probability, high-consequence events," Brennan noted. Certainty about where and when extreme impacts will occur is usually available only a few hours before those impacts begin— "well beyond an actionable timeframe for preparations." At 36-48 hours out, for example, a great deal of uncertainty is present, and so risk is communicated across a large geographical area even though the threat (e.g., high winds, storm surge) will ultimately only affect a small area.

Therefore, the tools used by the NHC to convey risk range from early forecasts to in-the-moment messaging. Brennan mentioned four tools that cover a range in terms of time and certainty. First, the 7-day tropical weather outlook is a probabilistic forecast that serves as an early alert that storm systems are forming. Messages are crafted to raise awareness of the storm formation, define a broad area of potential impact, and emphasize preparedness. The National Weather Service (NWS) coordinates the messages, with the aim of providing consistent information for use by the broadcast meteorology community, emergency managers (EMs) at the state and local levels, and other media.

The official storm forecast—which captures the forecast of the track, the intensity, and the size of the storm, as well as the cone that indicates the possible track of the center of the storm, is another such tool. This tool is also deterministic, she noted, and does not incorporate uncertainty.

A third category of tool is the "variety of probabilistic hazard-based products" that focus on specific, individual hazards associated with tropical cyclones—including storm surge, wind, flooding, rainfall, and tornadoes. These products do incorporate information about uncertainty, and they focus on the hazards present rather than predictions of the overall storm's track. At 3 to 5 days out, messaging becomes more focused on areas where the highest impact is anticipated; however, even at this point, "it's still too uncertain to get very specific about exact timing and magnitude." These messages also indicate how the risk is changing as the storm develops, she explained. This 3- to 5-day timeframe is often when decisions about preparations are made, and individuals and communities begin to take protective actions—especially when, for example, evacuation plans require a long lead time. At this point, Brennan noted, the NHC's messaging emphasizes the importance of following the advice of local officials and uses "rather severe wording" if needed, for example, when confidence is high that the event will be life-threatening.

Watches and warnings comprise the fourth type of tool used to communicate risk—in this case, about the risk of a particular hazard at a specific location. These messages become more detailed and vary depending on the time and location they cover. The focus here is more on the hazards than on the track or intensity of the overall storm, said Brennan. Each hazard requires a different response, and likely each location will be impacted at a slightly different time. She described how, during Hurricane Harvey, messaging from the NHC had to cover both the risk to the mid-Texas coasts of extreme wind and storm surge at landfall, and, several days later, the risk to the upper Texas coast of catastrophic flooding. Lead time is critical, and Brennan noted that the NHC uses watches and warnings in an effort to officially provide at least 36-48 hours of lead time. She echoed comments from previous discussants, including Pearson, Strickland, and LaDue, that communities often must begin to take protective action, such as evacuation, well before this mark.

Post-storm messaging is also critical, Brennan noted. This messaging might cover cleanup safety, heat risks, post-storm generator safety, and other topics. Raising these topics before the storm hits is important because reaching people once they are affected can be very difficult.

Alongside forecasts, messaging tools in and of themselves are critical to communicating risk in the event of a tropical cyclone. Brennan referred to "discussion products," which are designed to help forecasters convey information about uncertainty and communicate "how the risk is changing as the storm evolves." The NHC also provides impact-based decision support briefings to federal and state agencies, including the Federal Emergency Management Agency (FEMA), and sometimes local officials. Social media channels, including livestreams, are also an important part of the communication strategy and often are active in advance of the NHC's media pool, which is more formal and often "more focused on the last couple of days before landfall."

Castle Williamsberg, the second speaker on the panel, discussed challenges to "translating risk communication research for practitioner use," particularly with the aim of using research in the social, behavioral, and economic sciences to modernize and improve the Tropical Cyclone Product Suite at the NHC.[1] He described work by the Weather Program Offices (WPO) Social Science Program at NOAA, which funded four complementary studies, designed with intentional overlaps and differences, to yield a body of social science research on the impact of risk messaging products.[2] WPO and NWS social scientists synthesized the results of these four studies and worked with the NWS Tropical Roadmap Team to ensure that findings and future research were conceptualized in ways that were "operationally relevant and could be used in practice." Practitioner partners include EMs, forecasters, local officials, and other decision-makers.

Williamsberg highlighted five of the key themes from these studies as framed in operationally relevant terms. First, probabilistic information helps people make decisions in the midst of uncertainty. Practitioners could briefly explain how to interpret probability information, rather than simplifying their message and leaving out uncertainty all together. Second, partners expressed a strong desire for more localized information, even when level of uncertainty about the local forecast is high. Third, different types of "timing information" were critically useful to partners making decisions about risk communication. This information included not only when hazards might begin or reach the highest impact, but also when they might end and their duration (understand the impacts of sustained elements—for example, the impact of sustained wind to bridge infrastructure). Fourth, in situations of more than 5 days of lead time, partners wanted more information about

---

[1] More information about Tropical Cyclone Products at the NHC is available at https://www.nhc.noaa.gov/productexamples/ and https://www.nhc.noaa.gov/cyclones/.
[2] More information about the Weather Program Offices Social Science Program at NOAA is available at https://wpo.noaa.gov/social-science/.

the forecast models and forecast confidence. Linked to this were changes, suggested by partners, to NOAA's Tropical Cyclone Product Suite that would help "optimize the extraction of key information from textual and graphical products to further encourage message transmission." Finally, partners placed a high value on summary products— "key message products." A single dashboard or landing page that brought together "the entire ecosystem of tropical products" could help with message transmission.

Williamsberg then highlighted four challenges to translating research into actionable items for practitioner use that emerged from the synthesis of the studies' findings. The first challenge is understanding when researchers have enough information to ensure that study findings are usable by the practitioners. The second challenge involves addressing operational challenges and ensuring that the findings are "operationally relevant" (i.e., findings that are practicable by EMs, local officials, forecasters, and other partners). In collaboration with the NWS's Tropical Roadmap Team and with meteorologists, the WPO group developed findings that were beneficial to those practitioner partners and to "help accelerate their research-to-practice process." The third challenge relates to how best to share the findings among the intended audience, including meteorologists, EMs, local officials, and other decision-makers. The WPO also developed the System for Public Access to Research Knowledge (SPARK).[3] In addition, Williamsberg said, the WPO is developing a story map that will help make findings available to practitioners so that they, in turn, can be more attuned to what their own audiences and partners know or are focused on. The last challenge identified is how best to track the successful transfer of knowledge and the benefits associated with it. This effort would involve studying how practitioners incorporate research information into their work, learning how best to provide practitioners with research findings in usable forms, and tracking their usage of this new knowledge (see Porter et al., 2024).

Gina Eosco, also with NOAA's WPO, addressed the question of how to evaluate the effectiveness of risk communication. Risk communication is "inherently multi-sector," she noted, and therefore, to be effective, requires leveraging partnerships across the academic, public, and private sectors both in delivering messaging and in determining what kind of an impact such messaging makes. The first step, she noted, is establishing how success is measured: in this case, by looking at the impact of messaging during past events. To this end, WPO is developing the Societal Data Insights Initiative (SDII), a social science data infrastructure that integrates and synthesizes social and meteorological data, which enables users to gain insights into the societal effects of various products and services. Eosco emphasized that this endeavor is collaborative.

One challenge to the work of researching societal impact is gathering large, generalizable samples that more closely represent the public audience being served by these products. Eosco highlighted the importance of longitudinal research and event-based research, such as the studies outlined in Chapter 2. Partnering with

---

[3] More information about the SPARK platform is available at https://wpo.noaa.gov/empowering-open-science-unveiling-wpos-system-for-public-access-to-research-knowledge-spark/.

private-sector companies that develop weather apps to embed a survey in their app might, she suggested, yield a broader and more representative sample.

An important topic of research is how the public acts, and when, in response to specific products, Eosco said. Data on this topic are not available; "we need more evaluation capacity, and we need varying types of data to help answer these questions." In this, too, she suggested, a collaboration among public, private, and academic partners—who could share data (e.g., commercial, financial, retail)—that might yield better insights into how people respond to risk communication and where gaps might exist.

Another challenge is present in the idea that "the publics and partners make better-informed decisions when we communicate uncertainty," a theme raised repeatedly in earlier sessions. Eosco noted a fundamental disagreement: research shows that communicating uncertainty and probabilistic information leads to the best decision-making practices on the part of the public and partners. "Respected partners," on the other hand, consistently advocate for a simple message and deem probabilistic information either not useful or confusing. A fundamental disagreement around the effectiveness of risk communication, and how best to understand it, occurs if research says one thing and partners do not believe it, Eosco said. She stressed that the resolution to this question does not lie in determining which side is right or wrong; she notes there are likely limitations on both sides: bias and sample challenges on the research side, and heuristics that influence understanding on the public side. The more productive question, Eosco argued, is how the WPO might better understand the gap and why it exists, and how to "find that space of understanding." Trusted relationships between the public and private sector already support this approach, she noted. Eosco ended her presentation by explaining that NOAA and NWS are undertaking a paradigmatic shift in communication models, from a "one-size-fits-all" approach to a more personalized or localized model that addresses specific and unique needs. This effort requires knowledge of what those unique needs are, she explained, which happens over time and through strong partnerships.

Communicating uncertainty and probabilistic information to partners was the focus of the question-and-answer session, which centered on a query from Robby Goldman, an audience member, who wondered about examples of such communication done simply or in actionable ways. Eosco responded that all products are based on probabilistic information, even if that is not explicit in the outward-facing message, and that there are many ways to communicate uncertainty effectively. She noted that several messages and products used by the NHC incorporate uncertainty already. She pointed to Communication Assist Techs—CATs—that help people gain a better sense of whether their area will, for example, experience a dangerous storm surge.

Brennan added that products vary on their explicitness about the probabilistic information that underpin all of them; when uncertainty is not foregrounded, this is often an effort to translate risk into "a clearer, actionable message." Products that make probabilistic information explicit are, she noted, sometimes intended primarily for use by EMs and other experts who may be more interested in the specific details as they make decisions. "The goal is to put all of that probabilistic information out there for people to exploit, but not necessarily to put it all in the public-facing products," Brennan concluded.

MESSAGING TECHNOLOGY WALKTHROUGH

The second panel of the session involved demonstrations of new and emerging technologies in the risk communication field. Mike Gerber, Wireless Emergency Alert Expert and Meteorologist, NWS, the first speaker, described the many steps involved in distributing Wireless Emergency Alerts (WEAs) as used in the NWS Office of Dissemination and other alerting entities. WEAs are activated only during hazards that pose a great threat to life and property. In the tropical storm category, these include warnings for hurricanes and typhoons, as well as extreme wind. WEAs are template-based, with specific details filled in. There are both English and Spanish templates, as well as shorter and longer message lengths (90-character and 360-character templates). These messages are received by almost every cell phone.

WEA alerts involve cross-sector collaboration. The alerting process begins with a message originated by official alerting authorities, including more than 1,600 federal authorities, state agencies, territorial agencies, tribal governments, and local authorities. Using third-party authoring software, officials convert the message to common alerting protocol (CAP) format, which is based in XML and serves as the international standard format among alerting technologies. This CAP message is then sent to the Integrated Public Alert and Warning System (IPAWS), a CAP message aggregator run by the Federal Emergency Management Agency (FEMA).[4] At IPAWS, the CAP message is converted to another format—Commercial Mobile Alert for C-Interface (CMAC), which is specific to wireless alerts. The message in the CMAC format is attached to a polygon that indicates the geographical alert area. From here, the message moves from public agencies to private corporations within the wireless industry, who map the alert polygon onto their own network topology to identify appropriate cell towers from which to broadcast to the WEA. Phones receiving this message do not automatically display it; instead, they use device-based geofencing (DBGF), a process by which the phone compares its location, if known, with the polygon, to determine whether or not the user is actually in or close to the targeted area. If the user is inside this area, the phone displays the alert. This cuts down on over-messaging, as when, for example, the cell towers reach a larger area than the polygon. Gerber concluded by noting that, also in an effort to avoid over-messaging and desensitization to alerts, the NWS is focusing on streamlining messaging and using an impact-based warning approach to target affected areas.

Brock Aun, Vice President of Communication and Public Policy, HAAS Alert, then presented on Safety Cloud, a digital alerting delivery system that delivers re-

---

[4] More information about the FEMA IPAWS can be found at www.FEMA.gov/IPAWS or by inquiry to IPAWS@fema.dhs.gov.

al-time messages to drivers, built by HAAS Alert.[5] HAAS Alert has partnered with navigation apps WAZE and Apple Maps, which incorporate Safety Cloud into their own platforms and alert drivers to crashes, emergency vehicles, work zones, and other road hazards. Delivering alerts in this way gives drivers a small bit of advance warning, which, Aun said, can drastically reduce the odds of a crash in response to these various hazards. Aun pointed to a 2021 study by researchers at Purdue University that showed that, compared with the light bars and sirens found on most emergency vehicles, the digital warning helped reduce hard braking next to a roadside incident by 80 percent. The study estimated that only about 30 percent of drivers received the alerts, but even this minority was enough to influence behavior of most drivers. This particular message delivery platform offers several technical benefits, Aun noted: it aggregates information for automakers, it integrates into already-existing cloud code, it is a cross-platform solution to message delivery, and it has been integrated with IPAWS in certain areas to translate IPAWS data into real-time messaging. Safety Cloud is "aggregating the universe of roadway hazards and providing a single, authoritative stream for automakers." This effort involves building partnerships with automakers, who can incorporate this single system into various other connective systems already in place within the car. Safety Cloud integrates with more than 50 platforms already in vehicles on the road, including fire trucks, thanks direct work with emergency vehicle manufacturers to integrate Safety Cloud into platforms already in use. Aun noted that, throughout this effort, the main goal is not to provide every driver with a single platform, but rather, to connect the multitude of platforms already in use, making an "interoperable network of safety."

Aun concluded by mentioning questions that have been raised particularly around customization of alerts and messages. Customization—particularly, messages to drivers about specific hazards—is tricky, he explained, because, while messaging efforts should be led by federal agencies, other agencies and decision-makers have insights and want to have input as well. How best to alert drivers, and what to tell them in any given situation, are complex questions that highlight opportunities for further collaboration between the private and public sectors.

Philip Mai, Senior Researcher and Co-Director of the Social Media Lab, Toronto Metropolitan University, next spoke about his work with Anatoliy Gruzd, Professor and Co-Director of Research of the Social Media Lab, Toronto Metropolitan University, on developing social media research tools, such as dashboards and other visualizations, to further social science research on online misinformation. "Communalytic," a portmanteau of "community" and "analytics," is one such tool: "a no-code computational social science research tool for studying online communities and discourse." This suite of data collection modules collects publicly available data from social media websites such as Reddit, X, YouTube, and others. The tool allows for multiple approaches to analyzing the data, doing what Mai calls "social listening" to see how information flows between users.

---

[5] More information about Safety Cloud is available at https://www.haasalert.com/.

One area of focus is the coordination of crisis communication via social media. Coordination of this kind has become increasingly complex for several reasons, Mai noted. These reasons include the proliferation of new social media platforms, a scattered audience, a dearth of local news reporters, and the public's declining trust in traditional news sources, which as a result are much less powerful and present than they once were. Mai and Gruzd have focused on understanding how communicators get their message across in this difficult information environment, especially if the message is urgent or otherwise timely.

One challenge to crisis communication via social media is platforms' use of algorithmic filtering—that is, filtering what content individual users see and thus shaping their access to information—which dampens the likelihood of any single message going viral. Companies can pay to boost their messaging within this system, which also shapes what individual users see. Furthermore, Mai noted, social media systems are designed to privilege emotionally charged messages, and if an alert does not fit that criterion, it will not get shared as widely. Other challenges include misinformation generated by artificial intelligence (AI) and perpetuated by deep fakes; the disappearance of platforms and the resulting loss of audience; low levels of trust in information disseminated via social media; and, finally, the vulnerability of the system to state actors who weaponize it for their own ends (e.g., exaggerate, downplay, or otherwise misreport on a hazard or risk in order to discredit authorities' responses). Mai and Gruzd called for further consideration about how much weight should be given to information found online and also emphasized the importance of supporting traditional media, which serves as an important alternative source for information.

The brief discussion that followed focused on the question of whether alerting drivers to various hazards in real time would cause people to panic and thus cause more accidents. Gerber responded that different types of messages for different hazards tend to have different effects. What action the driver should be asked to take will vary based on the particular hazard. Customization is possible, and the host platforms—Waze, Apple Maps, etc.—will shape the message the driver actually sees. Therefore, he noted, standardizing as much of the messaging as possible from Safety Cloud, before it goes out into the multiple delivery systems, is important.

RISK COMMUNICATION INNOVATIONS AND NEW FRONTIERS IN TROPICAL CYCLONE COMMUNICATION: PRIVATE SECTOR

The last panel of the day featured three presentations about innovations on risk communication in the private sector. Mike Chesterfield, Vice President, Weather Presentation and Data Visualization, The Weather Channel, shared some of the work being done by the Weather Channel around risk communication in the production of what it calls "immersive mixed reality"—a new technique in hazard visualization that uses hyper-realistic video imagery to depict a forecasted scenario. "We [are] able to show what the future's going to look like in a video product." Showing video content was deemed important, Chesterfield noted, because studies found that visuals, including video, graphics, and images, increased the perception of risk, and that video in particular also lowered perceived uncertainty. In a way, he noted, these products "predict the future" in their depictions of "believable simulations" and in conjunction with on-camera meteorologists' commentary. The presence of the on-camera meteorologist is important, he explained, both for the expertise they bring and to cue the audience to the fact that the video they are seeing is not real footage, but a visualization. These videos are meant to provide important context to forecasts otherwise shown only on a map, Chesterfield explained. One such product, Surge-FX, debuted in 2018, shows the NHC's reasonable worst-case scenarios during an event (Figure 6.1). The message accompanying this product, and other such products, is that it is not a forecast, but a visual description of what people should be preparing for. Anecdotal evidence shows that this product and approach could influence how likely viewers are to take action in the event of an evacuation order or similar mandate, said Chesterfield, as shown in a recent in-house survey.

Flood-FX is another recent Weather Channel product that similarly depicts reasonable worst-case scenarios, in this case, by translating two-dimensional video into three-dimensional imagery. This product takes real footage of an area under threat and applies a simulation over it. These two products are intended to help viewers attend more closely to the forecast: "we hear all the time from on-camera meteorologists . . . [that] there is nothing more frustrating than having the forecast correct but people not listening," Chesterfield noted.

Surge-FX

FIGURE 6.1 A depiction of the Weather Channel's Surge-FX.
SOURCE: Presented by Mike Chesterfield on February 6, 2024, created by the Weather Channel.

One important element of Flood-FX, Chesterfield said, is its capacity for "hyperlocalization." Localized visualizations are a central goal going forward, he noted, describing how perhaps in the future, a product would allow users to enter their address and call up visualizations of what the forecast or various worst-case scenarios might look like in their front or back yard. "There is really no better way, in my opinion, to get people to react than actually show them what their future may actually look like."

Another innovation Chesterfield highlighted is "metahumans, or personal weather assistants." These are digital humans who would deliver personalized warnings to users—speaking in a way the individual understands, giving hyperlocal information, and using their name, for example. These products are, he said, somewhat controversial, but they do represent the capacity of this technology.

Micah Berman, Lead Project Manager, Android Platform Safety, Google, described Google's work to develop an earthquake early warning system (EWS) called Android Earthquake Alerts System (the System) for the Android phone platform, which is installed on about 3 billion devices worldwide.[6] The System is a detection and distribution system, supplemental to the national EWS and other government warning systems, that uses the accelerometer already in these devices to detect and model an earthquake's magnitude and epicenter in real time. This detection function is paired with distribution of alerts over a low-latency delivery network that is point-to-point internet protocol (IP) based. Berman explained that, at present, a bimodal model delivers one of two alerts, depending on the situation of the individual user. The first alert is a "be aware" alert, while the second is a "take action" alert. The two strands of action alert were developed to comply with official guidance set by multiple countries: a specific "drop, cover, and hold" instruction and a more general "protect yourself" instruction. These alerts are available in any of the languages supported by Android, Berman explained. They also are time sensitive and can be updated as the situation evolves.

Clicking through on an alert, users will see cached information about next steps. This feature is especially important in situations where power and connectivity may be unavailable. Information detected by ShakeAlert can also be pushed to the top of Google searches, so that details about the earthquake picked up by Android devices can appear in a regular search more quickly than information provided by official sources.

Berman noted that user feedback has been "extremely positive" and that areas of growth remain, foregrounding two questions: How to make earthquake warnings more actionable? and How to expand this capability to cover other types of hazards? Other possibilities for expansion might involve integrating with other capabilities already built into the phone, connecting people with information even if they are not in the locale (e.g., to allow them to learn more about loved ones who might be at risk), or giving a map and route in the case of evacuation.

---

[6] More information about the Android Earthquake Alerts System is available at https://crisisre sponse.google/android-alerts/.

John Lawson, Executive Director, AWARN Alliance, shared the alliance's work in using ATSC 3.0, the NEXGEN TV transmission standard, to improve the delivery of emergency alerts and messaging.[7] The warning system built on ATSC 3.0 technology differs from earlier ones, Lawson noted, distinguishing it from the Emergency Alert System (EAS) and WEA. The infrastructure underpinning ATSC 3.0 is TV towers, which are more resilient than cell towers. A battery-powered wi-fi dongle allows older television models to receive this signal and transmit over Bluetooth and wi-fi to other devices, he explained.

Lawson noted four important benefits of this delivery mode: the television signal provides geotargeting, which means that alerts—embedded in the television signal—will be received by a device that "knows where it is;" this locational information can enable targeted messaging. ATSC 3.0 also allows for rich media messaging that goes beyond text message alerts. Third, this technology enables most devices to be "woken up" if they are off but not unplugged; Lawson noted that this feature is very controversial. The final benefit is that ATSC 3.0 is the first and only internet protocol (i.e., IP)-based broadcast system for television. meaning it uses the IP for all types of information transmission.

In an ideal scenario around an "imminent alert," the alert authority would issue a message that would be picked up by the station and distributed via geotargeting to individuals within the warning polygon. Users could then interact with this information, whether to dismiss or learn more: "We want to give consumers optionality." Lawson noted that "interoperability" among alerting authorities and broadcasters is critical, explaining that, ideally, the alerting authorities and the broadcaster would at least enter into a memorandum of understanding "in terms of what they'll push out." Certain messages issued by alerting authorities would, ideally, go out "without anybody in the station touching it" rather than leaving the decision about whether the alert gets broadcast up to the news official.

At present, ATSC 3.0 is deployed across South Korea, Lawson said. In the United States, broadcasters are currently reaching around 75 percent of households that have televisions with an ATSC 3.0 signal. Outreach is critical to raising awareness of the existence of the NexGen TV/ATSC 3.0 signal and its potential for emergency messaging. AWARN Alliance developed some prototypes for EMs to use, and then conducted roundtables in which they could discuss the technology's use. Broadcasters involved in the roundtables expressed strong interest in this technology. Currently, a pilot project in Washington, D.C., is under way. However, national leadership is currently lacking, said Lawson, but is fully necessary for a "widescale deployment." These public-private relationships are critical for such a large-scale endeavor. Private solutions are in development as well, Lawson said, describing a start-up in which he is involved that would sell receivers that would bring advanced alerts, as well as a "disaster channel," a 24-hour broadcast that would be programmed by AI drawing on public domain content.

---

[7] More information about ATSC 3.0 is available at https://www.atsc.org/nextgen-tv/.

The question-and-answer session opened with a query from Jen Henderson about Eosco's idea that social science researchers might partner with companies to use their apps and other technologies to collect data via surveys or other tools. Lawson responded that social science research is central to their work; Berman echoed Lawson's enthusiasm for partnering with academic researchers. He noted that partnerships between large organization and researchers are sometimes not easy, but worth figuring out how to achieve.

An audience member offered two comments about "manufactured visuals." First, viewers might develop a skewed sense of their safety because they do not realize that what they are seeing is not real and, second, trauma-informed communication practices advise against messaging that makes impacts explicit and that repeatedly names or describes impacts—which the Weather Channel's products seem to be doing. Chesterfield responded that the Weather Channel takes great care to explicitly remind viewers that they are watching a simulation, in part by combining the visuals with narration that emphasizes this point, and to discourage imitators on social media from rebroadcasting images or videos of people who do put themselves in danger.

The final question came from Alex Lamers, NWS, who asked about the types of datasets that the NWS could gather and share that inform development of future innovations. Lawson noted that the roundtables revealed that "interoperability" was critical. Berman wondered whether, beyond earthquake detecting and alerting, other needs exist that the technology could address. He added that challenges around risk communication are not necessarily due to the data available from government sources; rather, they arise from gaps in the understanding of what users need and want, and what will compel them to take action. His concern, he said, is "less about the trigger and more about the experience that we built around that, and how we can make sure that is as effective and as integrated as it can be."

Chapter 7

New Approaches to Unmet Needs: Communication for the Whole Community

Clear communication rests on attention to both the crafting of the message and the needs of the target audience, a theme that was evident throughout the afternoon session of the second day, which included a keynote address by Wändi Bruine de Bruin, Provost Professor of Public Policy, Psychology, and Behavioral Science, University of Southern California, two panels, and a demonstration discussion of a prototype product. Ann Bostrom, moderator, listed the goals of the session, which were discussing "current and emerging strategies, barriers, and challenges for communicating uncertainty and probabilistic information," highlighting "unmet needs in communities at risk from tropical cyclones," and illuminating "potential solutions to meet those needs in the context of communication."

KEYNOTE SPEECH:
JARGON, TECHNICAL LANGUAGE, AND PLAIN LANGUAGE

Bruine de Bruin gave the day's keynote speech, on what Bostrom, in her introduction, summed up as the question of "how to be clear." In her talk, Bruine de Bruin stressed the important role of clear, accessible language in effective emergency messaging. While experts often spend a lot of energy ensuring that a message is correct and accurate, "message wording is often an afterthought." They tend to use complex language that is useful for communicating with other experts in their field but is not always accessible to the general public. Complex messages can be difficult to understand and off-putting, Bruine de Bruin explained. Therefore, "they may not work" and "they can put people's lives at risk."

Bruine de Bruin outlined a design process aimed at producing more effective messages. She illustrated the problems with technical or overly complex language that appears in such communications, and then described ways that experts who generate these messages might simplify their language in order to improve communication.

Social science research has illuminated approaches to making public messaging more effective, Bruine de Bruin said. Designing messages in advance of a crisis can help improve clarity and increase confidence in their effectiveness. She outlined four steps that comprise an effective design process. First, clearly identify recommendations. Second, explore why people may not follow those recommendations, perhaps by interviewing or surveying the target audience. Third, design a message based on the findings of that research to address the reasons why people do not follow the recommendations. Finally, test whether the message improves individuals' understanding of risk and inclination to protect themselves.

To illustrate the challenges that experts face when creating clear and jargon-free messaging, Bruine de Bruin described a project on climate change communication on which her team participated with the United Nations Foundation and the Intergovernmental Panel on Climate Change (IPCC), a group of climate science experts who create reports and other types of messages to be shared with policymakers, practitioners, and the general public. When the IPCC scientists were asked to identify the terms that would be central to climate change communications, two words stood out: mitigation and adaptation. These terms are central to behavior change, Bruine de Bruin noted: mitigation, as used by the scientists, refers to "the things that you can do to reduce your impact on climate change." Adaptation refers to "the things you can do to protect yourself against the climate change that is already happening."

Her team then conducted interviews with members of the general public, asking them to rank how easy the terms were to understand on a scale of one to five, and whether they could define the terms in their own words. The results of these interviews showed that "mitigation" was perceived as not easy to understand and that people often struggled to define it in their own words—sometimes confusing it with other words. "Adaptation" was rated relatively easy to understand, "but just because people think a term is easy to understand doesn't mean they define that term in the same way that the experts do." Definitions of the term "adaptation" given by interviewees centered on another meaning of that term: turning a book into a movie. Interviewees were then given the technical definition of the words and asked to suggest simple wording; their suggestions, Bruine de Bruin said, confirmed that it is truly possible to talk about these complex ideas in accessible ways. She then summarized the suggestions: mitigation became "actions we take to stop climate change from getting worse," while adaptation became "actions we take to protect against climate change." Terms used in risk communication about cyclones may also cause confusion among public audiences, Bruine de Bruin noted. Although the terms have not been systematically tested, she provided anecdotal evidence that people may find "shelter in place," "storm surge," "cone of uncertainty," and "watch/warning" confusing. Confusion around these terms can potentially distract from recommendations about actions one can take to protect oneself.

Simplifying language is important and possible. Bruine de Bruin outlined several suggestions drawn from research on effective risk communication. The first is that messages are written for a seventh-grade reading level or lower; the

Flesch-Kincaid readability test is one such measure. Writing in simple language also involves avoiding jargon, using short words—one-to-two syllable words common in everyday language—and short sentences. Feedback from the target audience is also important in determining whether the message effectively facilitated understanding and prompted behavior change. Bruine de Bruin concluded by noting that social scientists can work alongside tropical cyclone experts in gathering data about the effectiveness of messaging through surveys, interviews, and randomized experiments.

The discussion that followed began with a conversation about why jargon is used in messaging. Micki Olson, Senior Risk Communication Researcher and Project Manager, Emergency and Risk Communication Message Testing Laboratory, SUNY University at Albany, speculated that it might be a marker of credibility or expertise—and thus, a signal that their words should be heeded. Bruine de Bruin noted that some experts also resist using simple language because they worry about losing nuance or precision. However, she continued, if the goal of communicating is to "improve understanding in a particular community, you need to use the words of that community." She has found that when experts read excerpts from interviews with the target audience, they see more readily why and how confusion around complex, technical language arises. Bostrom wondered whether a tension ever existed in audiences composed of experts and members of the public—particularly where experts perceive technical language as having better precision. Bruine de Bruin responded that even for audiences of highly educated people, research shows that using clear, simple language still tends to work best. "It's only when you communicate with your own expert community that it may be important to use that complex language." She added, "If using precise terms is absolutely necessary, also describe the meaning in clear, everyday language so as to be as clear as possible."

COMMUNICATING UNCERTAINTY AND PROBABILISTIC INFORMATION ABOUT TROPICAL CYCLONE TRACKS, TIMING, AND SEVERITY

The panel that followed took up many of the same concerns and concepts introduced by Bruine de Bruin in her keynote speech: representing the complexity of information in a way that is simple for the intended audience to understand accurately. Lace Padilla, Assistant Professor, Khoury College of Computer Sciences, Northeastern University, spoke first, presenting research on hurricane visualizations, particularly on the different biases people have around understanding graphics that represent intensity, trajectory, and size of the storm. Padilla reported on a recent study in which her team compared respondents' biases and understanding of one such graphic, the cone of uncertainty, to an "ensemble technique" generated by utilizing the same underlying forecast model that created the cone of uncertainty graphic, making small perturbations to the model, and then sampling from the perturbation space. Each one of those lines represents one of those samples (Figure

7.1). The team first showed respondents a cone of uncertainty and asked them to use a rating scale to estimate how much damage that a specific location—a hypothetical off-shore oil rig—would incur. The ensemble technique graphics are based on the same forecast model that underpins the cone of uncertainty but are made by making small perturbations to the model; that is, samples are taken from this perturbation space and represented with individual lines that, together on the map, "show the distribution of the possible trajectories very intuitively." Lines clustered together indicate where the storm is more likely to pass, while areas with fewer lines or no lines—even within the cone of uncertainty—show where the storm is less likely to pass. This often looks "a little bit like a normal distribution" with some standard deviation. This ensemble technique can be used to show different points of time within the forecast, Padilla noted.

The results of this comparison showed two significant differences between the cone of uncertainty and the ensemble technique. First, Padilla reported, respondents tended to read the cone of uncertainty as a "danger zone," estimating high levels of damage inside the cone and much lower levels outside of the zone. This misconception was joined by a second: that a smaller cone of uncertainty means less damage. The ensemble technique, on the other hand, seemed to help

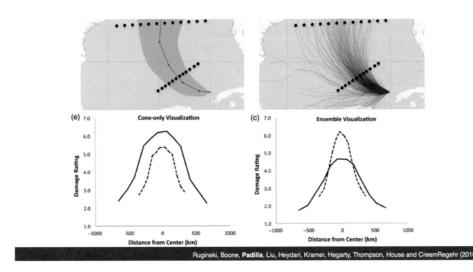

Ruginski, Boone, **Padilla**, Liu, Heydari, Kramer, Hegarty, Thompson, House and CreemRegehr (2016)

FIGURE 7.1 Different visualizations of the cone of uncertainty using the cone-only technique (right) and the ensemble technique (left). Mean damage ratings shown as a function of distance from center for each visualization for the 24-hour (dotted lines) and 48-hour (solid lines) timepoints.
SOURCE: Presentation by Lace Padilla on February 6, 2024; created by Ruginski et al. (2016). Reprinted by permission of Taylor & Francis Ltd.

respondents intuit a distribution based in probabilistic data and estimate damage levels with a more gradual decline across the space.

Padilla noted two key themes around understanding drawn from these findings: "intervals create conceptual categories" and "convention misalignments cause errors." In the case of the first theme, she explained that people tend to read lines or other spatial designations in graphics as meaningful delineations of important categories; here, respondents read the edges of the cone of uncertainty as stronger, more meaningful delineation than they really are. "The location of the exact boundary, set by the visualization designer" that delineates the cone of uncertainty is not as inherently meaningful as other, similar graphical delineations, she noted. The two sides of that boundary (inside and outside the cone) are not sharply different in terms of potential damage, although the public reads them as such.

The second theme is more general: cone of uncertainty graphics follow different rules than conventional maps. Specifically, the cone of uncertainty is an area designated by visualization designers and does not directly correspond to geographical distance the way most cartographic images do. "Pixel distance of that cone does not mean distance on a map; it means increased uncertainty." People tend to associate the size of the cone with the geographical size of the storm, read the image as such, and consequently, plan as such. This tendency is "very, very hard to override," even when people do know that the cone indicates level of uncertainty. Different displays can exacerbate this tendency if they show what should be smooth gradients as bands, which are often mistakenly read as categories, noted Padilla.

The ensemble technique disambiguates various parts of a storm that people assume are part of the cone of uncertainty visualization, said Padilla. It incorporates the path, showed with the various lines as described above. Uncertainty is represented with color, and size with a circle overlaying the image. This technique is an attempt to communicate clearly about all the elements that people often attribute to the cone of uncertainty.

Jessica Hullman, Ginni Rometty Associate Professor of Computer Science, Northwestern University, spoke next about the challenges to making probabilistic information accessible to the public and offered some strategies for increasing effective communication in that quarter. She began by noting that simply communicating uncertainty is not enough to guarantee that such information will be used appropriately by end-users. However, probabilistic language often leads to high levels of variance in terms of audience and interpretation, while graphics such as the cone of uncertainty often lead to biases such as those that underpin the categorical errors described by Padilla. "It's not that there are no good ways of representing uncertainty," Hullman noted; rather, probability itself is part of the challenge. People struggle to understand such an abstract, hypothetical way of presenting information, and often (mis)read probabilistic information as deterministic.

The first error Hullman mentioned is "deterministic construal errors," wherein people misread visual representations of probability as representation of various deterministic attributes of the storm (i.e., size, location). "As-if optimization" is a second type of error, wherein people suppress uncertainty in favor of a simple answer, whether that is a single location, a specific number, or otherwise. One expression of as-if optimization is rounding: for example, a person hears that there is a 22 percent likelihood of flooding in their area, and they assume that, because it is below 50 percent, they will be safe. Another example of as-if optimization is a fixation on the mean; even the most nuanced visualizations of distribution, including frequency formats, usually make visible what the central tendency is, and readers tend to focus on that information and use it to suppress uncertainty.

Uncertainty suppression is very difficult to avoid, Hullman noted. Her research shows that it is more likely to happen when people are under stress, dealing with a lot of information, and looking for a simple answer. Representing probabilistic information in ways that resist uncertainty suppression is complicated, she noted. "No matter how hard you try to design something that will force people to internalize uncertainty, it can be incredibly hard, because these tendencies to suppress uncertainty are so pervasive."

Hullman then outlined several successful strategies for making probability visible while resisting bias and variance (e.g., audience, interpretation) as much as possible. In general, these strategies involve making uncertainty visible. One approach presents draws from the distribution one wishes to show over time, rather than summarizing the information in a single, static graphic. This approach can often help people to interpret this information more intuitively. For example, hypothetical outcome plots refer to the use of probabilistic animation, "taking draws from the joint distribution we want to display and visualizing these frames as an animation" and making it more difficult to fixate on a single central number and easier to intuit information about probability. Hullman then showed an example of a quantile dotplot, a static "frequency-based representation of probability distribution function," and noted that giving people "this metaphor that probability is actually just the frequency" can improve their reasoning and decision-making. Although this approach offers real benefits to describing probabilistic information in terms of frequency, it can also backfire, Hullman noted. In one study, she investigated probability represented on various plots and found that even when participants did not have specific numbers (e.g., means), they had the same interpretations as if they had, because they were using the visual distance of the various data points to derive a deterministic reading and ignore the uncertainty.

A third approach to helping readers intuit uncertainty uses multiple maps and visualizations of several scenarios with simple language that explains their relationship, such as a narrative that helps readers understand that all of these scenarios could happen. This narrative approach might also include visualizations of scenarios that are highly likely, those that would be a little bit surprising, or those that are not very likely but would be catastrophic if they do occur. This approach involves determining how much visual emphasis to put on each scenario, and experimenting with readers to understand how much probability they would assign based on visualization conventions.

The discussion that followed built on the panelists' comments, starting a question from Cassandra Shivers-Williams, Social Science Deputy Program Manager, National Oceanic and Atmospheric Administration, Weather Program Office, about strategies to change or redirect common heuristics that often underpin uncertainty suppression. Hullman emphasized that frequency framing does work well, which involves showing multiple scenarios or multiple "draws" from the same probability distribution separately. These visuals can be animated or shown as a set of static images that represent multiple possibilities. With such approaches, people are not asked to interpret the size of an outlined area as probability, which can be very difficult for them to do. Padilla added that training people to read visual information differently is very difficult; therefore, if the goal is to change responses, then annotating a familiar image will not work—the visual information itself would have to be changed.

Ann Bostrom and Andrea Schumacher noted the importance of making and testing prototypes as part of the process of developing graphics that communicate uncertainty. They presented Hullman and Padilla with a prototype demonstration of an early-phase product that was developed by atmospheric scientists for research purposes and not for official use in risk communication. Schumacher explained that this exercise, in which Hullman and Padilla commented on the prototype, is part of an effort to bring social science into the design process earlier, rather than applying social science to a product that has already undergone years of development and is near completion. The prototype product is meant to express "the maximum wind speed exceedance values for the next 5 days" in the event of a hypothetical hurricane. It consists of two graphics: one that represents the wind speeds that are most likely and another that represents a "reasonable worst-case scenario" (Figure 7.2).

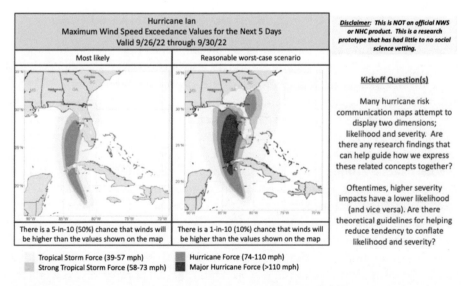

FIGURE 7.2 A research prototype designed by atmospheric scientists depicting the likelihood of various wind thresholds being exceeded over the next 5 days for Hurricane Ian (2022). SOURCE: Presented and created by Andrea Schumacher on February 6, 2024.

Padilla responded that the general public would likely focus on the visuals and ignore the textual annotations, or perhaps not understand them. Thus, people will likely come away with the misconception that the graphic is showing the track of the hurricane rather than estimates about wind speed. She also noted the likelihood of a categorical misperception, wherein people would interpret the brighter color as denoting where wind will most affect people— "Well, if I'm not there, then maybe I'm okay"—rather than probabilistic information. Schumacher explained that the graphic shows the impact area, but that, even so, it is helpful to understand that this information might be misconstrued as a cone of uncertainty, given the prevalence of that image.

Padilla noted the difficulty in communicating wind force speeds in a way that an individual can understand in a personal way. "It could be useful to remap the categories to people's individual experiences" to help them to better understand the potential impacts for them and choose appropriate responses. Such an approach might, for example, use colors to indicate low risk (i.e., speeds that affect unstable structures) versus high risk (i.e., speeds that affect all structures).

Hullman added that readers will likely place too much weight on the boundaries drawn to delineate areas in which these specific scenarios might play out: "Are you in the hurricane force wind or not? It's not exactly this line." With the current prototype, she explained, people would likely place equal weight on the two scenarios rather than reading them as two probabilistic outcomes. Indicating uncertainty remains critically important and might be communicated by showing people several scenarios that are highly likely and narrate these as such. Schumacher concluded the session by noting that this dialogue highlights the importance of co-development.

EXAMPLES OF ACCESS AND FUNCTIONAL NEEDS

The session's final panel featured presentations on reaching two specific audiences. Joseph Trujillo Falcón, Graduate Research Assistant, Cooperative Institute for Severe and High-Impact Weather Research and Operations, NOAA, & Bilingual Meteorologist, MyRadar, presented on communicating about emergencies in other languages than English, and Sherman Gillums, Jr., Director, Office of Disability Integration & Coordination, FEMA., presented on communicating with people with disabilities and other access and functional needs.

The stakes of communicating to multilingual populations are very high: the NWS has linked language inequity to fatalities since the 1970s, Trujillo Falcón asserted. With 69 million people in the United States who speak a language other than English, the stakes are only becoming higher: "it's not a matter of if, but when this becomes more consequential in our emergency communication systems" in the future. The monolingual emergency communication system—still in place today— is a barrier for many in the immigrant and multilingual populations and can limit understanding and effective decision-making. Trujillo Falcón recounted what inspires his work now: his experience as the only Spanish-speaking meteorologist at a radio station in a small town in Texas during Hurricane Harvey, which hit 2 weeks into his job. Witnessing the impacts of language vulnerabilities firsthand—including confusion of listeners who did not understand weather-related English information and instructions—allowed Trujillo Falcón to observe how people engage differently with information when it is in their dominant language, he explained.

"Language inaccessibility" is thus a challenge for non-English speakers in the United States, Trujillo Falcón noted. Making information more accessible brings its own challenges. Translation of specific terms can be difficult when there is no official standard definition of the term in English (Trujillo-Falcón et al., 2021, 2022). Another challenge lies in regional variations and cultural differences in Spanish that can skew meaning, he explained. For example, in Puerto Rican Spanish, "resaca" means "rip current," but in other parts of Latin America, "resaca" means "hangover;" so educational materials released by NOAA did not communicate the same thing in all places. Various agencies' inability to communicate consistently in Spanish for the variety of Spanish-speakers means that, for some, the hazard is not clearly or correctly portrayed, and risk is not properly understood, explained Trujillo Falcón.

Undocumented populations face particular challenges, not only around language accessibility but also in responding to perceived environmental risks, Trujillo Falcón explained, and policies around immigration can seriously complicate risk communication. People must weigh the risk of staying put in the face of a dangerous weather hazard or evacuating but getting arrested if identified as an undocumented immigrant. A 100-mile zone from the coast or international borders is the jurisdiction of the U.S. Immigration and Customs Enforcement (ICE), which gives officials the right to stop and search people suspected of illegal immigration without a warrant in this zone.[2] Trujillo Falcón recounted how, during Hurricane Harvey, President Donald J. Trump ordered ICE to keep immigration checkpoints open but local officials countered that order and kept the checkpoints closed. Even then, however, large groups within immigrant communities did not evacuate. Under the current administration, checkpoints are ordered to be closed during times of evacuation, a decision that, Trujillo Falcón noted, prioritizes saving lives, regardless of an individual's immigration status.

A third challenge that arises around risk communication for multilingual populations—especially Spanish-speaking people across North and South America—is cultural and community differences in understanding of what weather might bring what impacts. Messaging could be tailored to the specific community, in terms of not only translation but also social context. Because of context, information might be either readily understood or confusing, and attending to this difference is important when crafting effective messages. Trujillo Falcón illustrated this point by noting that someone from Puerto Rico will be familiar with hurricanes, while someone from another area—Perú, for example—will have never experienced

---

[2] More information about the 100-mile zone is available at https://www.aclu.org/know-your-rights/border-zone.

one. Thus, explaining what a hurricane involves might be redundant to a Puerto Rican, while comparing a weather event to a hurricane might be meaningless to a Peruvian.

Trujillo Falcón closed his presentation with a discussion of a grant project that he is leading at NOAA, which supports AI initiatives at the NWS to develop "neural adaptive translation software" that helps create translations of English messages into multiple languages, including Spanish, Vietnamese, and Mandarin. This AI produces automated translations that can be reviewed by forecasters, lessening their workload, especially in hectic times. So far, Trujillo Falcón reports that the NWS has had "great success," not only in communicating key messages but also in providing context that might be necessary in that particular setting: "We're able to bring people along from step one and say, this is what a hurricane hazard is. This is what you need to do. These are the recommendations that are involved in this. Let's break it down from the very first step." This software also integrates with geographic information systems (GIS) to identify the location of multilingual communities in the United States and to identify possible future collaborations.

Gillums then spoke about his work as FEMA's statutorily established disability coordinator. He noted that, from his perspective, the largest factor in risk communication is not the message so much as how people tend to see circumstances in a light that is most favorable to them, leading them to make decisions based on that optimistic bias. This tendency aligns with uncertainty suppression, discussed by Padilla and Hullman in the previous panel. Gillums shared that, because of an injury sustained while serving in the Marine Corps, he joined the 61 million people in the United States who identify as, or are regarded as, disabled. As with language accessibility, the stakes were very high around disability and decision-making during disasters in the past. "The reality was, if you were disabled or of advanced age, you were more likely to die in Hurricane Katrina than most other survivor populations."

Gillums highlighted the difference between safety and certainty that often shapes calculations that individuals make as they decide whether to heed alerts and evacuate. For many people, for example, the lack of safety inherent in evacuation is foreseeable unsafety, whereas whether a weather threat will actually impact them is not so predictable. Using his own experiences in the wildfires around San Diego on Labor Day weekend of 2005, Gillums described the difficulties around evacuation and other protective actions that people in the disability community face. He recounted his own process of rationalizing whether to stay at home or evacuate after hearing the voluntary evacuation order on television. Within "the window of uncertainty suppression"—when individuals have time to think before the threat hits—Gillums contemplated whether he should leave his house for a local football stadium where evacuees were advised to relocate. Despite having the intellectual capacity to think through the options, "in that moment, it became less about what was most safe, and more about what outcome seemed most certain for me." He noted that he did not feel safe about either option, but he felt more certain that if he stayed home and nothing happened, he would be safer than if he evacuated to the football stadium that he imagined would resemble the dire shelter conditions in Louisiana just a month earlier. For a time, it looked like the fire might reach him, and Gillums worried that he had made the wrong decision. Even in this moment, however, he knew that he might be unsafe at home, but he believed that his routine yet complex needs would be ignored in the stadium. "It wasn't about whether I'd be burned alive. It was about whether anyone would understand my needs were important to my day-to-day living." And the answer, he felt strongly, was no. He emphasized that for an individual living with a disability, their home is their sanctuary and place of peak autonomy and empowerment. Asking them to leave that sanctuary in anticipation of an event is an enormous burden because they may find themselves in an environment "where the uncertainty and potential for harm are profound to a point where it could have permanent implications." For a person with a disability, "the world is unsafe all the time." Therefore, the process of deciding whether to exchange one threat (staying home) for another (leaving home) can be very challenging.

Drawing on Mileti, Gillums related how people process decisions to his own experiences and work (Figure 7.3). The decision-making process starts with the message contents —what an individual is being told, he explained. In his own case, the information that reached him did not make the fire seem too dire; although he received the order to evacuate, he processed this information alongside the various reports "in a light most favorable" to himself.

FIGURE 7.3 A socio-behavioral model of warning response.
SOURCE: Presentation by Sherman Gillums, Jr. on February 6, 2024; adapted from work presented by Jeannette Sutton, PhD, at the Washington Partners in Emergency Preparedness webinar on Evidence-based guidance for effective alerts and warnings: The Warning Lexicon, November 28, 2023.

The second aspect of decision-making is how the message is sent: what language is being used, what associations are made (e.g., the meaning of a "bomb cyclone" related to a tropical storm in California to laypersons). "Message clarity is in the eye of the beholder," Gillums noted, and, therefore, front-end stakeholder inclusion during message development is key. Working with the target audience to develop the message, and then to further test the message, can improve the message's inclusivity.

Perception of the message is perhaps the most important aspect of decision-making, Gillums said. How do individuals understand the message? Do they believe it? Is it relevant to their circumstances, or could it be? Gillums explained that FEMA, in anticipation of a forecasted storm, reviews population data and other information to "triangulate ourselves into seeing a risk profile" and to better understand the individual needs of the communities it is trying to reach and serve who fall within the profile's context. For example, when a tornado hit the historic city of Selma, Alabama, near the Alabama School for the Deaf, in Talladega, Alabama, Gillums' team noted that the state needed to immediately secure American Sign Language (ASL) interpretation for the governor's addresses to the public. Relationships between agencies and communities are critical to ensuring that messaging is crafted to reach the intended group.

The final aspect of decision making—the response—can be described as either "moving or milling," explained Gillums. Whether people take action is influenced not only by the messaging, he explained, but also by who is making the decision (e.g., when fathers as decision-makers tend to stay, whereas mothers as decision-makers tend to go). Clear, effective communication involves a certain level of cultural competency: understanding who is likely to be participating in the decision, who is leading discussions, and who is likely to encounter the information. Gillums also stressed that messaging needs to be personalized to be effective. To illustrate this point, he recounted a conversation with a woman living on Sanibel Island who evacuated during Hurricane Ian and lost some neighbors who did not. She explained that, although the neighbors received the message about the risks and the order to evacuate, and although they "sensed danger," they did not believe the message was directed at them. The message did not feel personal, and the risk did not seem different from that of storms they had survived in the past.

Gillums concluded by noting FEMA's ongoing efforts to improve risk communication in this area, which include being more visible ahead of a storm by building relationships with communities, gaining a sense of the community and its needs, raising awareness, and helping communities prepare ahead of time. FEMA also aims to use language that speaks more fully to the disability community—for example, to indicate how to access transportation rather than simply giving an order to evacuate. Again, success depends on having strong connections with community stakeholders to more effectively strategize about communication.

The following discussion opened with a question from Sutton, the moderator, about whether specific words—especially jargon—are difficult to translate from one language to another using AI or within the disability community. Trujillo Falcón highlighted the word "trough," which can be translated as the place where pigs eat in some non-English languages. However, in areas where hurricanes are common, such as Puerto Rico and the Caribbean, a lot of good terminology already exists. Therefore, he explained, it is important to attend to regional differences in weather that correspond to differences in terminology. Working with certified translators is also very important, he noted, to better understand how dialectical varieties work and how best to communicate to different populations. Gillums noted that, on the one hand, people may not understand terms for impacts with which they are unfamiliar (e.g., lake effect snow), but, on the other hand, familiarity can also lull people into a false sense of security: they know what is coming and so, perhaps, miss details about the specific threat at hand. He further commented that "accessibility" means many things to many people and therefore no longer has consistent meaning.

A second question, from Schumacher, raised the idea that calculations around evacuation and other protective actions might be as much about resources (i.e., a person's perception of how well they will be treated and whether their needs will be met at a shelter or other site) as they are about communication itself. Gillums reiterated that clear, accessible communication is essential. But so, too, are resources, especially as climate change makes places vulnerable in new ways; this situation poses a dual challenge in that places are not prepared for such hazards and people are under-resourced. Helping people to understand "what they're up against" is critical, he explained; even if they cannot easily access resources, knowing what is needed can help people find ways to prepare and protect themselves.

Micki Olson asked the third and final question of the discussion: What policy does your agency need to create or implement? Trujillo Falcón responded that believes an agency should clearly define translation. He noted that in their work to develop bilingual risk communications, agencies supported the concept of translation, but in a literal sense so that the English and Spanish messages lined up perfectly. This approach, however, does not reflect how bilingualism and translation work, Trujillo Falcón noted: many words cannot be technically directly translated. "If we are more concerned about translating the meaning of given words overall, we'd be able to move a lot quicker" in producing messages that are consistent and resonant within the target communities.

Gillums responded that he advocates for FEMA to adopt a proactive rather than a reactive posture. Proactivity means not waiting too long to prepare people or ask them to take action: "There becomes a point when there are diminishing returns to the communication [of risk and what actions to take] because people have no decision-making capacity." Gillums stressed that this point is often reached sooner than people realize, and that if a storm is anticipated, then the time to engage in some of the important preparedness work and decision-making has passed. He acknowledged that agencies might have good reasons for adopting a reactive posture. Nevertheless, preparedness before an event is on the horizon is essential to keeping people in the disability population—and in general—safe.

HIGH-LEVEL SUMMARY OF SESSIONS FIVE AND SIX

Risk communication is about both the content of the message and the people who receive it. This central theme from sessions five and six was summarized by Andrea Schumacher, Jeanette Sutton, and Ann Bostrom. Bostrom commented that effective risk communication involves understanding how members of the general public make meaning from visualizations, and then developing products that align with those conventions and practices. Several panelists stressed the importance of "talk[ing] with intended audiences to learn what they think" about terms, communications products, and other aspects of messaging. Schumacher noted that public-private partnerships offer a promising opportunity to conduct this research. The perspective of social scientists is valuable; they should be engaged to advise on reception issues and audience understanding early in the product development process. "Thoughtfully entangling" these disciplines are tricky, Schumacher noted, and "transitioning social science research to operations requires a thoughtful approach that includes co-development evaluation and really, an iteration of those processes." The discussions raised awareness of audiences who are not being listened to, and of the unmet, unseen needs of "hidden communities," including immigrant populations.

A second variation on the theme of attending to both content and reception was the concept of customization, localized messaging, or messaging otherwise specific to a time, place, audience, and/or moment in the event. Schumacher commented that different messages are needed at different times and for different events, but also at different points within those events. The technology demonstration provided an example of how new technologies might support increasingly localized messaging (whether personalized or geographically targeted).

Another variation on this theme in found in how the process of communication intersects with uncertainty, which can have a meteorological or a more personal dimension. Messaging that foregrounds uncertainty and probabilistic information often changes to more deterministic messaging over the life of a storm, Schumacher explained; however, even in the most deterministic-sounding products, probabilistic information still underpins the forecast. "There are no 100 percent forecasts." Thus, useful information will always come with uncertainty. She pointed to a tension revealed by the panelists between the data showing that "people make better decisions with uncertainty information" and research showing that simple messaging and plain language are better for communication. The solution here is not either/or, but rather the development of a range of messaging strategies. Schumacher reiterated that more research is needed to know when, why, and for whom various communications approaches are effective. Bostrom noted the challenges inherent in communicating uncertainty via visualizations, which are often misunderstood by the general public. The many examples provided by panelists revealed the difficulty of finding an "intuitive and heuristic way of processing uncertainty communications" that guides readers to an accurate understanding of the situation. "Convention misalignments" can contribute to misunderstanding of visual information, and, again, research is necessary to improve knowledge of how people understand uncertainty through a variety of visualizations. Evoking another type of uncertainty, Sutton recalled Gillums' comments about how and why people, in their decision-making, might prioritize certainty over safety.

Another, and related, major theme of sessions five and six was plain language. Bostrom highlighted the ongoing conversation about using simple language and avoiding jargon. Sutton noted the problem of "semantic satiation," or the way a word loses urgency and meaning with repetition. Schumacher, in her mention of the tension between uncertainty and simple language, highlighted the importance of a spectrum of different messaging strategies.

Finally, partnerships between the public and private sectors yield opportunities for more and better research, noted Schumacher. Sutton added that new technologies provide more avenues for information dissemination. Bostrom noted that partnerships support a diverse array of perspectives and approaches and echoed a discussant's suggestion for the collaborative inclusion of "different populations, different audiences, different decision makers, different technology sciences, and different social sciences."

Chapter 8
Implications for the Future

Planning committee member, Marshall Shepherd, thanked participants for their contributions to the workshop and introduced the final activity. He explained that the committee intended the roundtable discussion to consider the workshop as a whole, synthesizing the rich material covered over the previous activities. Shepherd and fellow committee member Brad Colman served as moderators, and the panelists were Julie Demuth, National Science Foundation (NSF) and National Center for Atmospheric Research (NCAR), Brock Aun, HAAS Alert Systems, Gina Eosco, National Oceanic and Atmospheric Administration (NOAA) Weather Program Office (WPO), Sherman Gillums, Jr., Federal Emergency Management Agency (FEMA), and Rebecca Morss (NSF).

ROUNDTABLE DISCUSSION: WORKSHOP RETROSPECTIVE AND IMPLICATIONS FOR THE FUTURE

Shepherd opened the discussion by asking the panelists to share what surprised them about the workshop. Several panelists mentioned the challenges posed by increasingly complex data, tools, and approaches—together with the need to maintain simplicity in a complex context. Aun observed that many of the challenges to communication center on the interrelationship of message, medium, and audience, which become more complicated as people have more sophisticated understandings of the content, as well as more communications tools. These challenges will become more complex, he noted, and more and more robust approaches and sophisticated tools must be developed to adequately address the complexity. "I think that it's an illusion to think that there is one message that we need to get out there and keep it simple." Demuth highlighted the tension between the desire for hyper-localized information and efforts to communicate uncertainty, especially in relation to longer lead times. Although the ability to deliver hyper-local messages is increasingly supported by both tools and data, such messaging may not be viable early on, when precise information is not necessarily available, and uncertainty is high. "Simple isn't easy," added Eosco, urging audience members not to conflate the two. Crafting a "parsimonious simple message" can remain that goal, even though in and of itself it is a complex task.

Collaboration—especially in service of understanding the unique needs of various audiences—was another theme of the discussion. Gaining a thorough understanding of audiences and forecasts is critical, Eosco noted, but also complicated and should be sought through collaboration: "I don't think any one of us can do it alone." Morss agreed, noting her appreciation of the diversity of perspectives present; and not only present, but needed in order to address complex issues. With that diversity comes disagreement, but "everyone is really working in the same direction," she noted, toward the overarching goal of improving communication to reduce impacts.

"Preparedness is the first response." In this observation, Gillums summed up another theme of the roundtable discussion and workshop in general. Talking to people well before a disaster is anticipated, let alone imminent, helps them to understand the kinds of decisions they might face and to process how they might respond and prepare accordingly. He highlighted the "paradox between safety and certainty" that people with disabilities often face in moments of evacuation, and stressed the difficulty of helping people, especially in this community, to understand when it might be appropriate to trade one kind of certainty for another version of safety.

Colman asked panelists what they would prioritize in making risk communication more inclusive. Gillums, echoing the comments on collaboration by Eosco and Morss, emphasized the importance of relationships and trust between agencies and communities. Truly listening to people "who are not going to have great things to say" is critical if communication is to improve, even though they might "tell you you're inadequate." If people are not participating in conversations, then the field should try to understand their barriers to doing so. Demuth stressed the need for a broader view of the various barriers to understanding risk and making decisions in response that the public faces, for example, the tradeoff that Gillums described or the tradeoff that employees of the candle factory in Paducah, Oklahoma, faced when they were told that they would be fired (certainty) if they left their stations to seek safety from a tornado (uncertainty). In addition, for some people, evacuating might result in loss of income, if not loss of a job.

Aun noted that his company strives to make risk communication more inclusive by engaging people from the disability community early in the design process. This strategy stems in part from a mindset that the company is designing for everyone and therefore must intentionally include people with a range of abilities, privilege, and access: "everyone is a permanent part of local communities." Gillums cautioned against seeing the disability community as a monolith and emphasized the diversity of needs within that community. Rather than focusing on disability, he urged listeners to focus on "differences in experiences."

Eosco raised the question of how to know when there is enough research to make changes to current systems, or recommend them to partners, in order to make systems more inclusive. She noted the usefulness of a "gaps analysis" of the research and development process, identifying what has been done and what is needed, and prioritizing the latter. In 2023, the Weather Program Office (WPO) focused on diversifying its portfolio of projects, she explained, and has considered risk communication in different contexts, such as the deaf and hard-of-hearing community, immigrant populations, pregnancy and heat, and other "unique situations." Morss echoed and emphasized the necessity of identifying gaps and actively working to address them, stressing the need to identify who should be included. She added that listening to people once they are at the table is essential, and this includes attending to the full "complexity of their situation" and also revisiting conversations over time, as people, technologies, and situations change. Demuth added that "this qualitative work with people in the communities is so essential," adding that "if we don't know—really richly and in the complex ways that [Morss and Eosco] are talking about—how people are processing what some of those issues are that they're facing, we can't design some of the . . . quantitative work that is trying to measure that."

Colman then asked panelists to share their takeaways from the workshop and their ideas for plans to move forward: "Where are we going from here?" Morss shared that her largest takeaway was the importance of bringing together diverse perspectives. She pointed to the recurring discussion about how a hazard fits into a person's particular world: how meteorologists, social scientists, developers of new technologies, communications experts, and others all approach the various challenges and opportunities differently. Having each group understand the others' perspectives and ensuring that the work of each is "contextually relevant" to the others is critical, she noted. Morss also highlighted the importance of toggling between "the real world" in all its complexity and research to develop understanding in more simplified contexts: "How do you take the questions from the real world and do research . . . or build systems to address them in a more focused way?" This effort requires collaboration across perspectives, and Morss said, "I think that we're finally at that point where there are enough people in the room" to address these complex questions. Echoing Morss, Eosco noted the importance of bridging research and practice and to ensuring that knowledge gained in research is effectively shared with and used by partners and the general public.

Referring to discussions about localization and personalization of information, Eosco stated a question that framed one of her takeaways from the workshop: "How do we create an agile forecast to meet the needs of every user while still maintaining an official forecast?" Aun added that personalization and localization are "the entire future that we are going towards." For him, this future entails not only customization opportunities and geographically tailored messaging, but also communication systems with feedback loops so that data can be gathered toward optimizing use, expanding accessibility, and boosting efficacy. Ideally, this system would be built with the aim of establishing a national approach based on best practices.

Demuth raised the idea of a "predictions challenge," evident in the tension between clear and accurate risk communication, on the one hand, and the extent to which events and impacts simply cannot be predicted or known, on the other hand. Much of the discussion around risk communication assumes the presence of knowledge, she said, but this is not always the case. Rapid intensification is one instance of a prediction challenge, as are co-occurring compounding hazards. Predicting impacts is important, but in the case of the extremes, mapping impacts onto the meteorology can be very challenging, she said, reiterating a point made by Jeff Lindner in his presentation the previous day. Demuth also reiterated Lindner's question about the role of the meteorologist and whether or not they are responsible for predicting impacts.

Gillums reminded the group that human beings are, and must be kept, visible and at the center of all of this work. In that vein, he urged attendees not to underestimate the importance of their own experiences, perspectives, emotions, and predictions about how they might react to certain situations or questions—and to leverage those insights as they conduct research and craft messaging.

Shepherd then asked panelists to comment on any gaps or significant opportunities for further work they observed over the course of the workshop. Eosco responded, offering four major touchpoints. First, there is no good way to evaluate the system of forecasts and impacts; with so many components within the system, how can an individual component be evaluated? Second, returning to the theme of localization versus predictability, she noted that many current tools are rooted in a one-size-fits-all mode of thought. One gap, then, is building new forecast systems that enable more agility and customization, which would involve learning the various needs of users. Third, when resources are limited, understanding and targeting hyper-focused audiences can be challenging. Finally, artificial intelligence (AI) should be considered a tool, and not a solution. Aun noted a gap between "the human element at the local leadership level" and the potential for new forms of technology that might help to enhance decision-making by local officials.

Several panelists offered other gaps in research, language, and perspective. The critical incident stress experienced by forecasters, emergency managers (EMs), broadcasters, officials, and community members is another gap in knowledge, Demuth observed. A broad frame of study might be useful here, she noted, and might explore whether these groups are at risk for moral injury when a lack of capacity causes them to be unable to protect the lives and livelihoods of the people they serve. Gillums observed that the pervasiveness of a "deficits perspective" constitutes a sort of gap. To this end, he advised, "assume what you're doing is already working to an extent," and voiced a hope that people will not underestimate how successful their work is and has been in helping people think through decisions during hazardous events. Information is trickling through and having a positive effect: "What you're doing is saving lives." Morss, offering a final response to this question, observed that the partnerships, learning, and conversations deepened over the course of the workshop will help to tackle the gaps mentioned above.

Colman noted that, although a great deal of uncertainty is inherent in weather information, other areas exhibit much more certainty. These areas include knowledge about communities, the built infrastructure, and the physical environment, which can be drawn on to mitigate or inform uncertainty about weather information, especially in the work of communicating information in personalized or localized ways. "Don't let the uncertainty piece necessarily be the driver of personalization," he commented.

An audience member shared the perspective of EMs who make and communicate about decisions around evacuation orders and other protective measures. They noted that panelists might learn a great deal from close study of the moments when a decision is made, attending to both how quickly EMs must make decisions and how deliberate they are in making those decisions. Communicating the decision-making process to the public has also been important in their community, the audience member explained: who is making decisions, who has the authority to do so, and what expert advice is informing the decision. Aun noted his own surprise at learning, through his work, how under-resourced first responders are, and how much is expected of them. Taking on more risk with fewer resources is an impossible task, he observed.

Bob Hershey, an audience member, wondered about the possibility of representing probability of damage and loss resulting from a storm in terms of monetary value. This exercise might be done ahead of time as a way to communicate how the cost of damage might increase or decrease depending on the path of the storm, he suggested. Gillums responded that the idea is interesting, but that people might still misunderstand the concept of probability that underpins those numbers. He added that uncertainty sometimes means the spread of probabilities is very large.

A third audience member asked about approaches to supporting partnerships—particularly because communities and local groups might already be overtaxed and under-resourced. Gillums noted that a FEMA remit is to assess local needs, which is shared by municipal, state, and tribal authorities often working with the federal government. He noted that FEMA's reactive position is problematic: resources are available, but many communities do not know what they need. This gap could be mitigated though ongoing relationships that help communities prepare for such events, but because FEMA becomes involved after an event, these relationships and pre-event preparedness work often do not occur, so communities cannot take full advantage of the federal resources available. "I don't think it's a resource problem, at least from the federal government side," he commented. Aun noted that federal funds to expand the HAAS Alert system and to provide for basic equipment for first responders are drawn from the same pool. Rather than compete for those limited resources, he said, "we had to partner with our customers to go to the federal government" and lobby for new funding sources.

Eosco, in response to the earlier comments about EMs, noted that partnerships could also enable real-time observation. Often research is conducted after the fact, which can yield different results than research conducted in the moment. She wondered about the possibility of joining a local office during an event—not to reduce people to research subjects, but to leverage a real-time opportunity to increase empathy and understanding around the pressures and responsibilities that decision-makers face in the moment. She emphasized the importance of academics going to EMs and conducting research in person and in their environment; she also noted that grants can include funding for EM travel. Using virtual tools to connect researchers and local officials in the moment might be a possibility, she noted, especially because the increased use of virtual tools as a product of the pandemic has made partnering more inclusive, reducing the need for physical travel.

WORKSHOP WRAP-UP

Bostrom concluded the workshop by celebrating the rich conversations over the course of the 2 days, and expressing hope that the workshop would have lasting positive effects on work to come. The work of the community was likewise inspiring, and she emphasized that many advances and successes that were brought to light in the various panels and discussions. Bostrom then briefly highlighted some of the main themes of the workshop. First, "preparedness is the best and first response" in a myriad of situations, including effective risk communication and building strong partnerships. Furthermore, she noted, research efforts can help in this area by facilitating cultural competence and supporting localization and personalization of preparedness.

Taking an earth systems approach—working across all the sciences—to improve risk communication and to learn from unprecedented and extreme weather events is an important second theme, Bostrom said. This approach involves developing effective ways to evaluate the success of the risk communication system and its constituent parts, and designing systems that have "dynamic population representation and inclusive feedback loops." The field could employ a national approach to evaluate how well the entire system is working. For example, feedback loops could elicit the decisions and experiences of the entire population, including hard-to-reach populations and communities, and provide feedback throughout the evolution of an extreme weather event. Another area of opportunity, Bostrom highlighted, centers on partnerships between researchers in various branches of science—meteorology, social sciences, computational science—and partners on the ground, as well as communities impacted by tropical cyclones and other hazards. Bostrom concluded her remarks by noting that the workshop itself served as the beginning point for rich conversations among people in many areas. These conversations illuminated gaps, opportunities, and successes, and Bostrom expressed hope that people would continue to forge relationships and build on the work done during the workshop.

References

Bowker, G. C., and S. L. Star. 2000. *Sorting Things Out: Classification and Its Consequences*. MIT Press.

Cova, T. J., D. Li, L. K. Siebeneck, and F. A. Drews. 2021. Toward Simulating Dire Wildfire Scenarios. *Natural Hazards Review* 22(3):06021003. DOI: 10.1061/(ASCE)NH.1527-6996.0000474.

Cutter, S. L. 2018. Compound, cascading, or complex disasters: What's in a name? *Environment: Science and Policy for Sustainable Development* 60(6):16-25. DOI: 10.1080/00139157.2018.1517518.

Demuth, J. L., R. E. Morss, G. Wong-Parodi, A. B. Schumacher, J. J. Alland, D. Smith, N. Herbert, and H. D. Walpole. 2023. Understanding people's evolving risk assessments and decisions during tropical cyclone threats: Design and implementation of a novel, longitudinal, real-time survey methodology. https://doi.org/10.25923/wk8q-k529

Garfin, D. R., R. R. Thompson, E. A. Holman, G. Wong-Parodi, and R. C. Silver. 2022. Association between repeated exposure to hurricanes and mental health in a representative sample of Florida residents. JAMA Network Open 5(6):e2217251. https://doi.org/10.1001/jamanetworkopen.2022.17251

Haney, T., J. R. Elliott, E. Fussell, D. Brunsma, D. Overfelt, and J. Picou. 2010. Risk, roles, resources, race, and religion: A framework for understanding family evacuation strategies, stress, and return migration. *In The Sociology of Katrina: Perspectives on a Modern Catastrophe*. Rowman & Littlefield.

Henderson, J., E. R. Nielsen, G. R. Herman, and R. S. Schumacher. 2020. A hazard multiple: Overlapping tornado and flash flood warnings in a National Weather Service Forecast Office in the Southeastern United States. *Weather and Forecasting* 35(4):1459-1481. DOI: 10.1175/WAF-D-19-0216.1.

Li, Y., Y. Tang, S. Wang, R. Toumi, X. Song, and Q. Wang. 2023. Recent increases in tropical cyclone rapid intensification events in global offshore regions. *Nature Communications* 14(1):5167. DOI: 10.1038/s41467-023-40605-2.

Loughe, A. F., S. Madine, and J. Mahoney. 2008. *A Lead-Time Metric For Assessing Skill in Forecasting the Onset of IFR Conditions.* National Oceanic and Atmospheric Administration Earth System Research Laboratory, Boulder, CO: 1-10.

Nielsen, E. R., G. R. Herman, R. C. Tournay, J. M. Peters, and R. S. Schumacher. 2015. Double impact: When both tornadoes and flash floods threaten the same place at the same time. *Weather and Forecasting* 30(6):1673-1693. DOI: 10.1175/WAF-D-15-0084.1.

Olson, J., C. Doyle, D. LaDue, and A. Marmo. 2023. End-user threat perception: Building confidence to make decisions ahead of severe weather. *Journal of Operational Meteorology*:95-109. DOI: 10.15191/nwajom.2023.1108.

Porter, M., R. Hernández, B. Checkoway, E. R. Nielsen, C. Williamsberg, G. Eosco, K. Christian, A. Morris, E. Grow Cei, K. Patelski, and J. Henderson. 2024. Expanding the concept of knowledge transition through social science research. *Bulletin of the American Meteorological Society* 105(4):E816-E824. DOI: 10.1175/BAMS-D-23-0310.1.

Ruginski, I. T., A. P. Boone, L. M. Padilla, L. Liu, N. Heydari, H. S. Kramer, M. Hegarty, W. B. Thompson, D. H. House, and S. H. Creem-Regehr. 2016. Non-expert interpretations of hurricane forecast uncertainty visualizations. *Spatial Cognition & Computation* 16(2):154-172. DOI: 10.1080/13875868.2015.1137577.

Thompson, R. R., D. R. Garfin, and R. C. Silver. 2017. Evacuation from natural disasters: A systematic review of the literature. *Risk Analysis* 37(4):812-839. DOI: 0.1111/risa.12654.

Trujillo-Falcón, J. E., O. Bermúdez, K. Negrón-Hernández, J. Lipski, E. Leitman, and K. Berry. 2021. Hazardous weather communication en Español: Challenges, current resources, and future practices. *Bulletin of the American Meteorological Society* 102(4):E765-E773. DOI: 10.1175/BAMS-D-20-0249.1.

Trujillo-Falcón, J. E., A. R. Gaviria Pabón, J. T. Ripberger, A. Bitterman, J. B. Thornton, M. J. Krocak, S. R. Ernst, E. C. Obeso, and J. Lipski. 2022. ¿Aviso o Alerta? Developing effective, inclusive, and consistent watch and warning translations for U.S. Spanish speakers. *Bulletin of the American Meteorological Society* 103(12):E2791-E2803. DOI: 10.1175/BAMS-D-22-0050.1.

Wanless, A., S. Stormer, J. T. Ripberger, M. J. Krocak, A. Fox, D. Hogg, H. Jenkins-Smith, C. Silva, S. E. Robinson, and W. S. Eller. 2023. The Extreme Weather and Emergency Management Survey. *Weather, Climate, and Society* 15(4):1113-1118. DOI: 10.1175/WCAS-D-23-0085.1.

Appendix A
Public Meeting Agendas

**COMMITTEE ON ADVANCING RISK COMMUNICATION
WITH DECISION-MAKERS FOR EXTREME TROPICAL CYCLONES**

**VIRTUAL | NAS Building, 2101 Constitution Ave. NW,
Washington, DC 20418**

**FEBRUARY 5, 2024
LECTURE ROOM**

10:30: **Welcome and Opening Remarks**
Ann Bostrom, University of Washington, Committee Chair

10:40: **SESSION 1—OPEN**
Communicating Risks of Atypical Tropical Cyclones: Lessons from Henri and Hilary

Moderators: Andrea Schumacher, NSF-NCAR, &
Ann Bostrom, University of Washington, Planning Committee Members

10:40: **Panel 1: Forecaster Perspectives**
Alex Lamers, National Weather Service Weather Prediction Center
Rose Schoenfeld, National Weather Service
Robbie Berg, National Hurricane Center

Panel 2: Research Perspectives
Roxane Cohen Silver, University of California, Irvine
Julie Demuth, NSF National Center for Atmospheric Research
Emma Spiro, University of Washington

12:00: Lunch

12:45: SESSION 2—OPEN
Risk Communication in Multi-Hazard Environments: Challenges and
Learning Opportunities from Compounding Hazards and
Cascading Impacts

Moderator: Marshall Shepherd, University of Georgia

Keynote Speaker
Jen Henderson, Texas Tech University

1:00: Panel:
Jason Senkbeil, University of Alabama
Rebecca Moulton, FEMA
Jeff Lindner, Harris County Flood Control District
Jessica Schauer, National Weather Service Tropical Cyclone
    Weather Services Program

1:50: Break

2:05: High Level Summary of Sessions 1 and 2

2:15: Transition to Breakout Rooms

2:40: SESSION 3—OPEN
Breakout Discussions: Applying Risk Communication Lessons from
Other Hazards to the Tropical Cyclone Context

Moderator: Jeannette Sutton, SUNY Albany

Room 1: Earthquakes
Richard Allen, University of California, Berkeley
Michele Wood, California State University, Fullerton
Sara McBride, USGS

Room 2: Extreme Heat
Micki Olson, SUNY, Albany
Olga Wilhelmi, National Center for Atmospheric Research
Peter Howe, Utah State University

Room 3: Flooding
Amanda Schroeder, National Weather Service
Rachel Hogan Carr, Nurture Nature Center

2:55: Report Back to Plenary
Moderator: Jeannette Sutton, SUNY Albany

3:15: Break

3:30: SESSION 4—OPEN
Risk Communication and Decision Making in Communities

Moderators: Craig Fugate, Craig Fugate Consulting LLC, &
Brad Colman, American Meteorological Society

3:30: Panel: Risk Communication Across Scales: Risk Communicators
in Communities

Drew Pearson, Dare County Emergency Management
Russel Strickland, Maryland Department of Emergency Management
Daphne Ladue, University of Oklahoma
Tom Cova, University of Utah
Jim Elliott, Rice University

Roundtable: Community Leaders and Community Action
Jeff Lindner, Harris County Flood Control District
Peyton Siler-Jones, National League of Cities (NLC)
Archie Chaisson, Lafourche Parish Government
Randy Reid, International City/County Management Association (ICMA)

4:45: High Level Summary of Sessions 3 and 4

4:55: Wrap Up and Plans for Day 2
Ann Bostrom, University of Washington, Committee Chair

5:00 END OF DAY 1

FEBRUARY 6, 2024
LECTURE ROOM

10:30: Welcome and Opening Remarks
Ann Bostrom, University of Washington, Committee Chair

10:35: Recap from Day 1

10:50: SESSION 5—OPEN
Practical Translation of Risk in the Public Arena

Moderators: Andrea Schumacher, NCAR,
Jeannette Sutton, SUNY Albany, &
Gabrielle Wong-Parodi, Stanford University

Panel 1: Risk Communication Innovations and New Frontiers in TC Communication: Public Sector

Mike Brennan, National Hurricane Center
Castle Williamsberg, FedWriters Supporting NOAA's Weather
    Program Office
Gina Eosco, NOAA Weather Program Office

10:50: Messaging Technology Walkthrough
Mike Gerber, National Weather Service
Brock Aun, HAAS Alert
Anatoliy Gruzd & Philip Mai, Toronto Metropolitan University

Panel 2: Risk Communication Innovations and New Frontiers in TC Communication: Private Sector
Mike Chesterfield, The Weather Channel
Micah Berman, Google
John Lawson, AWARN

12:30: Lunch

1:15: SESSION 6—OPEN
New Approaches to Unmet Needs: Communication for the Whole Community

Moderators: Ann Bostrom, University of Washington &
Jeannette Sutton, SUNY Albany

Keynote Speaker: Jargon, Technical Language, and Plain Language
Wändi Bruine de Bruin, University of Southern California

Panel 1: Communicating Uncertainty and Probabilistic Information about TC Tracks, Timing and Severity
Lace Padilla, Northeastern University
Jessica Hullman, Northwestern University

Panel 2: Access and functional needs
Sherman Gillums, Jr., FEMA
Joseph Trujillo-Falcon, University of Oklahoma

2:50:   **Break**

3:05:   **High Level Summary of Sessions 5 and 6**

3:15:   **SESSION 7—OPEN**
Workshop Retrospective and Implications for the Future

Moderators: Marshall Shepherd, University of Georgia &
Brad Colman, American Meteorological Society

Roundtable
Julie Demuth, NSF National Center for Atmospheric Research
Brock Aun, HAAS Alert
Sherman Gillums, Jr., FEMA
Gina Eosco, NOAA Weather Program Office
Rebecca Morss, U.S. National Science Foundation

4:15:   **Wrap Up and Closing Remarks**
Ann Bostrom, University of Washington, Committee Chair

4:30:   **Adjourn**

Appendix B
Committee Biographies

Ann Bostrom (*Chair*) is the Weyerhaeuser endowed Professor in Environmental Policy at the Evans School of Public Policy and Governance, University of Washington. Until 2007 she was Professor of Public Policy and Associate Dean for Research of the Ivan Allen College of Liberal Arts at Georgia Institute of Technology, and she co-directed the Decision, Risk and Management Science Program at the National Science Foundation (NSF) from 1999 to 2001. She studies risk perceptions, risk communication, and mental models of hazards: how people understand and make decisions under uncertainty about, for example, climate change, extreme weather, and earthquakes. Bostrom currently co-directs the NSF-funded Cascadia Coastlines and Peoples Hazards Research Hub and co-leads risk communication in the NSF Artificial Intelligence (AI) Institute for research on Trustworthy AI in Weather, Climate and Coastal Oceanography. Bostrom previously served as the task team co-lead for the National Oceanic and Atmospheric Administration's Science Advisory Board "Priorities for weather research" report. She is also a Fellow and former President of the Society for Risk Analysis, and recipient of its Chauncey Starr and Distinguished Educator Award. She is also a Fellow of the American Association for the Advancement of Science and an elected member of the Board of Directors of the Washington State Academy of Sciences. Bostrom received a Ph.D. in policy analysis from Carnegie Mellon University. She also received an M.B.A. from Western Washington University and a B.A. in English from the University of Washington. She co-chaired the National Academies of Sciences, Engineering, and Medicine consensus report on Integrating Social and Behavioral Sciences Within the Weather Enterprise and contributed to Communicating Science Effectively: A Research Agenda.

Dereka Carroll-Smith is a Postdoctoral Research Associate for the National Institute of Standards and Technology-Professional Research Experience Program at the University of Maryland College Park and a research meteorologist for the National Wind Impacts Reduction Program. Carroll-Smith also holds a joint appointment as a Program Coordinator and Adjunct Professor in the Department of Chemistry, Physics, and Atmospheric Sciences at Jackson State University, and as a Scientific Visitor at the National Center of Atmospheric Research (NCAR), where she conducts interdisciplinary research focusing on secondary tropical cyclone hazards, climate change, and associated societal impacts. While in graduate school, she received the David M. Knox endowment fellowship and the National Science Foundation Graduate Research Fellowship, which allowed her the freedom to explore her interdisciplinary interests. Carroll-Smith is a member of the American Meteorological Society, serves on the steering committee for the Significant Opportunities in Atmospheric Research and Science program at NCAR, and served as co-rapporteur of the tropical cyclone tornado section for the World Meteorological Organization's 10th workshop on Tropical Cyclones. Carroll-Smith received a B.S. in meteorology from Jackson State University, an M.S. in atmospheric science from Purdue University, and a Ph.D. in atmospheric science from the University of Illinois Urbana-Champaign.

Brad R. Colman is currently serving as President of the American Meteorological Society (AMS). Prior to this role he served as Director of Weather Strategy for Bayer/The Climate Corporation where he oversaw and guided the design and execution of the Bayer Enterprise weather programs. Before joining Bayer/Climate, Colman worked on a new Microsoft consumer weather service team to serve weather information across the entire Microsoft ecosystem. Previously, Colman had a diverse career with the National Oceanic and Atmospheric Administration (NOAA) where he worked at the National Weather Service's forecast office in Seattle, Washington; at NOAA's Environmental Research Laboratory; and as the Acting Director of NOAA's Meteorological Development Laboratory. Colman is a member and Fellow of the AMS, is a member of the Washington State Academy of Sciences, and is currently co-chair of NOAA's Science Advisory Board's Environmental Information Services Working Group. Colman received a B.S. in Earth sciences and mathematics from Montana State University and an Sc.D. in atmospheric sciences from the Massachusetts Institute of Technology. He currently serves on the National Academies of Sciences, Engineering, and Medicine's Roundtable on Macroeconomics and Climate Change and the Board on Atmospheric Sciences and Climate.

W. Craig Fugate provides senior-level advice and consultation in disaster management and resiliency policy through Craig Fugate Consulting LLC. Previously, he served as the Administrator of the Federal Emergency Management Agency (FEMA) and the Florida Emergency Management Director from 2001 to 2009. Fugate led FEMA through multiple record-breaking disaster years and oversaw the federal government's response to major events such as the Joplin and Moore Tornadoes, Hurricane Sandy, Hurricane Matthew, and the 2016 Louisiana flooding. He successfully managed the devastating effects of the 2004 and 2005 Florida hurricane seasons (Charley, Frances, Ivan, Jeanne, Dennis, Katrina, and Wilma).

Fugate holds a certificate as a paramedic from Santa Fe College in Gainesville, Florida. He also serves as a member of the National Academies of Sciences, Engineering, and Medicine's Gulf Research Program Division Committee.

Michael Lindell is an Emeritus Professor, Texas A&M University; Affiliate Professor, University of Washington Department of Urban Design and Planning; Affiliate Professor, Boise State University Department of Geosciences; and Affiliate Professor, Oregon State University School of Civil and Construction Engineering, and currently serves as a consultant on two hurricane warning and evacuation research projects funded by the National Science Foundation (NSF) and U.S. Army Corps of Engineers. He has conducted emergency management research and provided technical services to 40 different organizations in the public and private sectors and conducted research on topics ranging from surveys of disaster warning response to the development of an evacuation management decision support system. He also conducted a series of hurricane evacuation planning studies for the Texas Division of Emergency Management during his term as the Director of the Texas A&M University Hazard Reduction & Recovery Center. He has received awards from the International Sociological Association and the Human Factors and Ergonomics Society for his development of the Protective Action Decision Model, which summarizes research on human response to disaster warnings. Lindell received a Ph.D. in social psychology at the University of Colorado, Boulder, while working on the first NSF-funded Assessment of Research on Natural Hazards.

Andrea Schumacher is a Project Scientist in the Weather Risks and Decisions in Society research group at the National Center for Atmospheric Research (NCAR) in Boulder, Colorado. She works in the interdisciplinary space between atmospheric and social science, and her most recent research focuses on how information use, risk perceptions, and behavioral responses evolve in the days prior to a landfalling hurricane. Previously she was a Research Associate at the Cooperative Institute for Research in the Atmosphere at Colorado State University, where she was the Lead of the Societal Impacts of Weather and Climate Team, tropical cyclone (TC) forecast product developer, and satellite liaison. She has collaborated extensively with operational TC forecasters in the National Weather Service, especially on the topic of communicating TC wind hazards and probabilities to a variety of decision makers. Her work on the National Hurricane Center TC wind speed probability product earned her an Outstanding Achievement Award in Meteorology from the National Hurricane Conference and a Leadership Award from the Louisiana Emergency Preparedness Association. Schumacher received an M.S. in atmospheric science from Colorado State University.

Marshall Shepherd is the Georgia Athletic Association Distinguished Professor of Geography and Atmospheric Sciences at the University of Georgia and Director of its Atmospheric Sciences Program. Prior to academia, he spent 12 years as a scientist at the National Aeronautics and Space Administration Goddard Space Flight

Center and was Deputy Project Scientist of the Global Precipitation Measurement Mission. Shepherd is the host of The Weather Channel's Weather Geeks Podcast, is a senior contributor to Forbes Magazine, and has three TEDx talks on climate science and communication. Shepherd is a recipient of a Presidential Early Career Award for Scientists and Engineers, the Captain Planet Foundation Protector of the Earth Award, the 2019 AGU Climate Communication Prize, the 2020 Mani L. Bhaumik Award for Public Engagement with Science, and the 2018 American Meteorological Society (AMS) Helmut Landsberg Award. He is an elected member of the National Academy of Sciences, National Academy of Engineering, and the American Academy of Arts and Sciences and was the 2013 President of AMS. Shepherd received a B.S., M.S., and Ph.D. from Florida State University. He currently serves as a member of the National Academies of Sciences, Engineering, and Medicine's Board on Atmospheric Sciences and Climate.

Jeannette Sutton is currently an Associate Professor in the Department of Emergency Preparedness and Homeland Security at the University at Albany where she directs the Emergency and Risk Communication Message Testing Lab. Sutton has led research associated with natural, technological, and human-induced phenomena, with a focus on alerts and warnings over short messaging channels. She served for 6 years as the primary social scientist on the National Institute of Standards and Technology National Construction Safety Team Advisory Committee. Sutton received a Ph.D. from the University of Colorado Boulder and completed her postdoctoral training at the Natural Hazards Center. She previously served as cochair of the National Academies of Sciences, Engineering, and Medicine's workshop on Public Response to Alerts and Warnings on Mobile Devices.